Petroleum Science and Engineering

Petroleum Science and Engineering

Deepak Sharma

Petroleum Science and Engineering

ISBN 978-93-5111-417-8

Published in 2014 in India by

RANDOM PUBLICATIONS

4376-A/4B, Gali Murari Lal, Ansari Road
New Delhi-110 002
Phone : +91-11-43580356, +91-11-23289044
e-mail: randomexports@gmail.com, sales@randompublications.com, info@randompublications.com

Reprinted 2021

Type Setting by: Friends Media, Delhi-110089
Digitally Printed at: Replika Press Pvt. Ltd.

Preface

Energy is a key component to people's everyday lives; and a secure energy future requires a balance between environmental impact and affordable supply. Petroleum and geosystems engineers are able to address and solve important issues that will lead to energy security and thus are in high demand. Economic and environmentally safe production of petroleum resources requires creative application of a wide spectrum of knowledge. Petroleum engineering is a field of engineering concerned with the activities related to the production of hydrocarbons, which can be either crude oilor natural gas. Exploration and Production are deemed to fall within the upstream sector of the oil and gas industry. Exploration, by earth scientists, and petroleum engineering are the oil and gas industry's two main subsurface disciplines, which focus on maximizing economic recovery of hydrocarbons from subsurface reservoirs. Petroleum geology and geophysics focus on provision of a static description of the hydrocarbon reservoir rock, while petroleum engineering focuses on estimation of the recoverable volume of this resource using a detailed understanding of the physical behaviour of oil, water and gas within porous rock at very high pressure. The combined efforts of geologists and petroleum engineers throughout the life of a hydrocarbon accumulation determine the way in which a reservoir is developed and depleted, and usually they have the highest impact on field economics. Petroleum engineering requires a good knowledge of many other related disciplines, such as geophysics, petroleum geology, formation evaluation (well logging), drilling, economics, reservoir simulation, reservoir engineering, well engineering, artificial lift systems, completions and oil and gas facilities engineering.

Petroleum engineers increasingly use advanced computers, not only in analysis of exploration data and simulation of reservoir behaviour, but also in automation of oilfield production and drilling operations. Petroleum companies own many of the world's supercomputers. Petroleum engineers have a future full of challenges and opportunities. They must develop and apply new technology to recover hydrocarbons from oil shale, tar sands, and offshore oil and gas fields. They must also devise new techniques to recover oil left in the ground after application of conventional producing techniques. Since many petroleum companies conduct worldwide operations, the petroleum engineer may have the opportunity for assignments all over the world. Petroleum engineers must solve the variety of technological, political, and economic problems encountered in these assignments. These exciting technological challenges combine to offer the petroleum engineer a most rewarding career.

The present book deals with all the important dimensions of this subject. It is a valuable reference source for all those concerned with this subject.

—Deepak Sharma

Contents

1

Petroleum Geology

Petroleum geology refers to the specific set of geological disciplines that are applied to the search for hydrocarbons (oil exploration).

Sedimentary Basin Analysis

Petroleum geology is principally concerned with the evaluation of seven key elements in sedimentary basins:

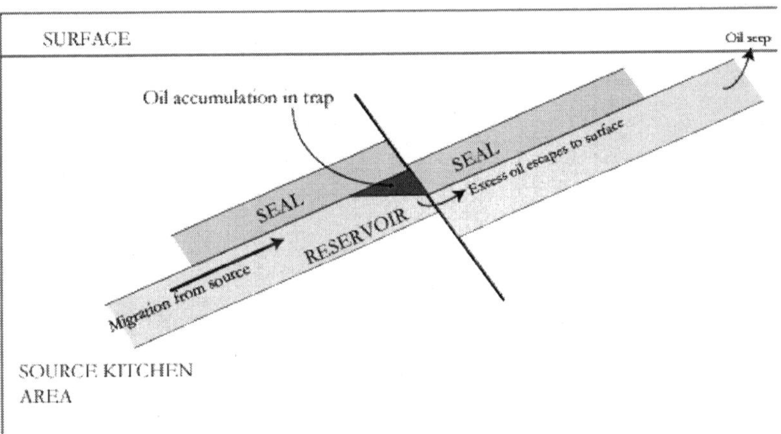

Figure: *A structural trap, where a fault has juxtaposed a porous and permeable reservoir against an impermeable seal. Oil (shown in red) accumulates against the seal, to the depth of the base of the seal. Any further oil migrating in from the source will escape to the surface and seep.*

- Source
- Reservoir
- Seal

- Trap
- Timing
- Maturation
- Migration.

In general, all these elements must be assessed via a limited 'window' into the subsurface world, provided by one (or possibly more) exploration wells. These wells present only a 1-dimensional segment through the Earth and the skill of inferring 3-dimensional characteristics from them is one of the most fundamental in petroleum geology. Recently, the availability of inexpensive, high quality 3D seismic data (from reflection seismology) and data from various electromagnetic geophysical techniques (such as Magnetotellurics) has greatly aided the accuracy of such interpretation. The following section discusses these elements in brief.

Evaluation of the source uses the methods of geochemistry to quantify the nature of organic-rich rocks which contain the precursors to hydrocarbons, such that the type and quality of expelled hydrocarbon can be assessed.

The reservoir is a porous and permeable lithological unit or set of units that holds the hydrocarbon reserves. Analysis of reservoirs at the simplest level requires an assessment of their porosity (to calculate the volume of *in situ* hydrocarbons) and their permeability (to calculate how easily hydrocarbons will flow out of them). Some of the key disciplines used in reservoir analysis are the fields of structural analysis, stratigraphy, sedimentology, and reservoir engineering.

The seal, or *cap* rock, is a unit with low permeability that impedes the escape of hydrocarbons from the reservoir rock. Common seals include evaporites, chalks and shales. Analysis of seals involves assessment of their thickness and extent, such that their effectiveness can be quantified.

The trap is the stratigraphic or structural feature that ensures the juxtaposition of reservoir and seal such that hydrocarbons remain trapped in the subsurface, rather than escaping (due to their natural buoyancy) and being lost.

Analysis of maturation involves assessing the thermal history of the source rock in order to make predictions of the amount and timing of hydrocarbon generation and expulsion.

Finally, careful studies of migration reveal information on how hydrocarbons move from source to reservoir and help quantify the source (or *kitchen*) of hydrocarbons in a particular area.

Major Subdisciplines in Petroleum Geology

Several major subdisciplines exist in petroleum geology specifically to study the seven key elements discussed above.

Analysis of Source Rocks

In terms of source rock analysis, several facts need to be established. Firstly, the question of whether there actually *is* any source rock in the area must be answered. Delineation and identification of potential source rocks depends on studies of the local stratigraphy, palaeogeography and sedimentology to determine the likelihood of organic-rich sediments having been deposited in the past.

If the likelihood of there being a source rock is thought to be high, the next matter to address is the state of thermal maturity of the source, and the timing of maturation.

Maturation of source rocks depends strongly on temperature, such that the majority of oil generation occurs in the 60° to 120°C range. Gas generation starts at similar temperatures, but may continue up beyond this range, perhaps as high as 200°C.

In order to determine the likelihood of oil/gas generation, therefore, the thermal history of the source rock must be calculated.

This is performed with a combination of geochemical analysis of the source rock (to determine the type of kerogens present and their maturation characteristics) and basin modelling methods, such as back-stripping, to model the thermal gradient in the sedimentary column.

Basin Analysis

A full scale basin analysis is usually carried out prior to defining leads and prospects for future drilling. This study tackles the petroleum system and studies source rock (presence and quality); burial history; maturation (timing and volumes); migration and focus; and potential regional seals and major reservoir units (that define carrier beds). All these elements are used to investigate where potential hydrocarbons might migrate towards. Traps and potential leads and prospects are then defined in the area that is likely to have received hydrocarbons.

Exploration Stage

Although a basin analysis is usually part of the first study a company conducts prior to moving into an area for future exploration, it is also sometimes conducted during the exploration phase. Exploration geology comprises all the activities and studies necessary for finding new hydrocarbon occurrence. Usually seismic (or 3D seismic) is shot and old exploration data (seismic lines, well logs, reports) are used to expand upon. Sometimes gravity and magnetic studies are conducted and oil seeps and spills are mapped to find potential areas for hydrocarbon occurrences. As soon as a significant hydrocarbon occurrence is found by an exploration- or wildcat-well the appraisal stage is set in.

Appraisal Stage

The Appraisal stage is used to delineate the extent of the discovery. Furthermore reservoir properties, connectivity, hydrocarbon type and gas-oil and oil-water contacts are determined to calculate potential recoverable volumes. This is usually done by drilling more appraisal wells around the initial exploration well. Furthermore some production tests may give insight in reservoir pressures and connectivity. While geochemical and petrophysical analysis gives information on the type (viscosity, chemistry, API, carbon content, etc) of the hydrocarbon and the nature of the reservoir (porosity, permeability, etc).

Production Stage

After a hydrocarbon occurrence has been discovered and appraisal has indicated it is a commercial find the production stage is initiated. This stage focuses on extracting the hydrocarbons in a controlled way (without damaging the formation, within commercial favourable volumes, etc). Production wells are drilled and completed in strategic positions. 3D seismic is usually available by this stage to target wells precisely for optimal recovery. Sometimes enhanced recovery (steam injection, pumps, etc) is used to extract more hydrocarbons or to redevelop abandoned fields.

Analysis of Reservoir

The existence of a reservoir rock (typically, sandstones and fractured limestones) is determined through a combination of regional studies (i.e. analysis of other wells in the area), stratigraphy and sedimentology (to quantify the pattern and extent of sedimentation) and seismic interpretation. Once a possible hydrocarbon reservoir is identified, the key physical characteristics of a reservoir that are of interest to a hydrocarbon explorationist are its bulk rock volume, net-to-gross ratio, porosity and permeability.

Bulk rock volume, or the gross rock volume of rock above any hydrocarbon-water contact, is determined by mapping and correlating sedimentary packages. The net-to-gross ratio, typically estimated from analogues and wireline logs, is used to calculate the proportion of the sedimentary packages that contains reservoir rocks.

The bulk rock volume multiplied by the net-to-gross ratio gives the net rock volume of the reservoir. The net rock volume multiplied by porosity gives the total hydrocarbon pore volume i.e. the volume within the sedimentary package that fluids (importantly, hydrocarbons and water) can occupy. The summation of these volumes for a given exploration prospect will allow explorers and commercial analysts to determine whether a prospect is financially viable.

Traditionally, porosity and permeability were determined through the study of hand specimens, contiguous parts of the reservoir that outcrop at the surface and by the technique of formation evaluation using wireline tools passed down the well itself. Modern advances in seismic data acquisition and processing have meant that seismic attributes of subsurface rocks are readily available and can be used to infer physical/sedimentary properties of the rocks themselves.

Underbalanced Drilling

Underbalanced drilling, or UBD, is a procedure used to drill oil and gas wells where the pressure in the wellbore is kept lower than the fluid pressure in the formation being drilled. As the well is being drilled, formation fluid flows into the wellbore and up to the surface. This is the opposite of the usual situation, where the wellbore is kept at a pressure above the formation to prevent formation fluid entering the well.

In such a conventional "overbalanced" well, the invasion of fluid is considered a kick, and if the well is not shut-in it can lead to a blowout, a dangerous situation. In underbalanced drilling, however, there is a "rotating head" at the surface - essentially a seal that diverts produced fluids to a separator while allowing the drill string to continue rotating.

If the formation pressure is relatively high, using a lower density mud will reduce the well bore pressure below the pore pressure of the formation. Sometimes an inert gas is injected into the drilling mud to reduce its equivalent density and hence its hydrostatic pressure throughout the well depth. This gas is commonly nitrogen, as it is non-combustible and readily available, but air, reduced oxygen air, processed flue gas and natural gas have all been used in this fashion.

Kinds of Underbalanced Drilling

There are several kinds of underbalanced drilling. The most common are listed below.

- Dry Air. This is also known as dusting. Here air compressors combined with a booster (which takes the head from the compressors and increases the pressure of the air, but does not increase the volume of air going down hole) are used and the only fluid injected into the well is a small amount of oil to reduce corrosion.
- Mist. A small amount of foaming agent (soap) is added into the flow of air. Fine particles of water and foam in an atmosphere of air bring cuttings back to the surface.
- Foam. A larger amount of foaming agent is added into the flow. Bubbles and slugs of bubbles in an atmosphere of mist bring cuttings back to the surface.
- Stable foam. An even larger amount of foaming agent is added into the flow. This is the consistency of a shaving cream.
- Airlift. Slugs and bubbles of air in a matrix of water, soap can or can not be added into the fluid flow of air.
- Aerated Mud. Air or another gas is injected into the flow of drilling mud. Degassing units are required to remove air before it can be recirculated.

Advantages

Underbalanced wells have several advantages over conventional drilling including:

- Eliminated formation damage. In a conventional well, drilling mud is forced into the formation in a process called invasion, which frequently causes formation damage - a decrease in the ability of the formation to transmit oil into the wellbore at a given pressure and flow rate. It may or may not be repairable.

In underbalanced drilling, if the underbalanced state is maintained until the well becomes productive, invasion does not occur and formation damage can be completely avoided.

- Increased Rate of Penetration (ROP). With less pressure at the bottom of the wellbore, it is easier for the drill bit to cut and remove rock.
- Reduction of lost circulation. Lost circulation is when drilling mud flows into the formation uncontrollably. Large amounts of mud can be lost before a proper mud cake forms, or the loss can continue indefinitely. If the well is drilled underbalanced, mud will not enter the formation and the problem can be avoided.
- Differential sticking is eliminated. Differential sticking is when the drill pipe is pressed against the wellbore wall so that part of its circumference will see only reservoir pressure, while the rest will continue to be pushed by wellbore pressure. As a result the pipe becomes stuck to the wall, and can require thousands of pounds of force to remove, which may prove impossible. Because the reservoir pressure is greater than the wellbore pressure in UBD, the pipe is pushed away from the walls, eliminating differential sticking.
- Formation damage Some rock formation have a reactive tendency to water. When drillmud is used the water in the drill mud reacts with the formation (mostly clay) and inheriently causes a formation damage (reduction in permeability and porosity) Use of underbalanced drilling can prevent it

Disadvantages

Underbalanced drilling is usually more expensive than conventional drilling (when drilling a deviated well which requires directional drilling tools), and has safety issues of its own. Technically the well is always in a blowout condition unless a heavier fluid is displaced into the well.

Air drilling requires a faster up hole volume as the cuttings will fall faster down the annulus when the compressors are taken off the hole compared to having a higher viscosity fluid in the hole. Because air is compressible mud pulse telemetry measurement while drilling (MWD) tools which require an incompressible fluid can not work. Common technologies used to eliminate this problem are either electromagnetic MWD tools or wireline MWD tools. Downhole mechanics are usually more violent also because the volume of fluid going through a downhole motor or downhole hammer is greater than an equivalent fluid when drilling balanced or over balanced because of the need of higher up hole velocities. Corrosion is also a problem, but can be largely avoided using a coating oil or rust inhibitors.

Oil Spill

An oil spill is the release of a liquid petroleum hydrocarbon into the environment, especially marine areas, due to human activity, and is a form of pollution. The term is mostly used to describe marine oil spills, where oil is released into the ocean or coastal waters. Oil spills may be due to releases of crude oil from tankers, offshore platforms, drilling rigs and wells, as well as spills of refined petroleum products (such as gasoline, diesel) and their by-products, heavier fuels used by large ships such as bunker fuel, or the spill of any oily refuse or waste oil. Another signifant route by which oil enters the marine environment is through natural oil seeps.

Oil spills can be controlled by chemical dispersion, combustion, mechanical containment, and/or adsorption. Spills may take weeks, months or even years to clean up.

Environmental Effects

The oil penetrates into the structure of the plumage of birds and animals, reducing its insulating ability, thus making the birds more vulnerable to temperature fluctuations and much less buoyant in the water. It also impairs or disables birds' flight abilities to forage and escape from predators. As they attempt to preen, birds typically ingest oil that covers their feathers, causing kidney damage, altered liver function, and digestive tract irritation. This and the limited foraging ability quickly causes dehydration and metabolic imbalances. Hormonal balance alteration including changes in luteinizing protein can also result in some birds exposed to petroleum.

Most birds affected by an oil spill die unless there is human intervention. Some studies have suggested that, even after cleaning, less than 1% of oil soaked birds survive. Marine mammals exposed to oil spills are affected in similar ways as seabirds. Oil coats the fur of Sea otters and seals, reducing its insulation abilities and leading to body temperature fluctuations and hypothermia. Ingestion of the oil causes dehydration and impaired digestions. Because oil floats on top of water, less sunlight penetrates into the water, limiting the photosynthesis of marine plants and phytoplankton. This, as well as decreasing the fauna populations, affects the food chain in the ecosystem. There are three kinds of oil-consuming bacteria. Sulfate-reducing bacteria (SRB) and acid-producing bacteria are anaerobic, while general aerobic bacteria (GAB) are aerobic. These bacteria occur naturally and will act to remove oil from an ecosystem, and their biomass will tend to replace other populations in the food chain. Oil may also cause the death of an animal by entering the animal's lungs or liver. The animal will then be poisoned by the oil. Oil also can kill an animal by blinding it. The animal will not be able to see and be aware of their predators. If they are not aware of other animals, they may be eaten.

Cleanup and Recovery

Cleanup and recovery from an oil spill is difficult and depends upon many factors, including the type of oil spilled, the temperature of the water (affecting evaporation and biodegradation), and the types of shorelines and beaches involved.

Methods for cleaning up include:

- Bioremediation: use of microorganisms or biological agents to break down or remove oil.
- Bioremediation Accelerator: Oleophilic, hydrophobic chemical, containing no bacteria, which chemically and physically bonds to both soluble and insoluble hydrocarbons. The bioremediation accelerator acts as a herding agent in water and on the surface, floating molecules to the surface of the water, including solubles such as phenols and BTEX, forming gel-like agglomerations. Undetectable levels of hydrocarbons can be obtained in produced water and manageable water columns. By overspraying sheen with bioremediation accelerator, sheen is eliminated within minutes. Whether applied on land or on water, the nutrient-rich emulsion creates a bloom of local, indigenous, pre-existing, hydrocarbon-consuming bacteria. Those specific bacteria break down the hydrocarbons into water and carbon dioxide, with EPA tests showing 98% of alkanes biodegraded in 28 days; and aromatics being biodegraded 200 times faster than in nature they also sometimes use the hydrofireboom to clean the oil up by taking it away from most of the oil and burning it.

- Controlled burning can effectively reduce the amount of oil in water, if done properly. But it can only be done in low wind, and can cause air pollution.
- Dispersants act as detergents, clustering around oil globules and allowing them to be carried away in the water. This improves the surface aesthetically, and mobilizes the oil. Smaller oil droplets, scattered by currents, may cause less harm and may degrade more easily. But the dispersed oil droplets infiltrate into deeper water and can lethally contaminate coral. Recent research indicates that some dispersants are toxic to corals.
- Watch and wait: in some cases, natural attenuation of oil may be most appropriate, due to the invasive nature of facilitated methods of remediation, particularly in ecologically sensitive areas such as wetlands.
- Dredging: for oils dispersed with detergents and other oils denser than water.
- Skimming: Requires calm waters
- Solidifying: Solidifiers are composed of dry hydrophobic polymers that both adsorb and absorb. They clean up oil spills by changing the physical state of spilled oil from liquid to a semi-solid or a rubber-like material that floats on water. Solidifiers are insoluble in water, therefore the removal of the solidified oil is easy and the oil will not leach out. Solidifiers have been proven to be relatively non-toxic to aquatic and wild life and have been proven to suppress harmful vapours commonly associated with hydrocarbons such as Benzene, Xylene, Methyl Ethyl, Acetone and Naphtha. The reaction time for solidification of oil is controlled by the surf area or size of the polymer as well as the viscosity of the oil. Some solidifier product manufactures claim the solidified oil can be disposed of in landfills, recycled as an additive in asphalt or rubber products, or burned as a low ash fuel. A solidifier called C.I.Agent (manufactured by C.I.Agent Solutions of Louisville, Kentucky) is being used by BP in granular form as well as in Marine and Sheen Booms on Dauphin Island, AL and Fort Morgan, MS to aid in the Deepwater Horizon oil spill cleanup.
- Vacuum and centrifuge: oil can be sucked up along with the water, and then a centrifuge can be used to separate the oil

from the water - allowing a tanker to be filled with near pure oil. Usually, the water is returned to the sea, making the process more efficient, but allowing small amounts of oil to go back as well. This issue has hampered the use of centrifuges due to a United States regulation limiting the amount of oil in water returned to the sea.

Equipment used includes:

- Booms: large floating barriers that round up oil and lift the oil off the water
- Skimmers: skim the oil
- Sorbents: large absorbents that absorb oil
- Chemical and biological agents: helps to break down the oil
- Vacuums: remove oil from beaches and water surface
- Shovels and other road equipments: typically used to clean up oil on beaches.

Prevention

- Secondary containment - methods to prevent releases of oil or hydrocarbons into environment.
- Oil Spill Prevention Containment and Countermeasures (SPCC) program by the United States Environmental Protection Agency.
- Double-hulling - build double hulls into vessels, which reduces the risk and severity of a spill in case of a collision or grounding. Existing single-hull vessels can also be rebuilt to have a double hull.

Environmental Sensitivity Index (ESI) Mapping

Environmental Sensitivity Index (ESI) maps are used to identify sensitive shoreline resources prior to an oil spill event in order to set priorities for protection and plan cleanup strategies. By planning spill response ahead of time, the impact on the environment can be minimized or prevented. Environmental sensitivity index maps are basically made up of information within the following three categories: shoreline type, and biological and human-use resources.

Shoreline Type

Shoreline type is classified by rank depending on how easy the garet would be to clean up, how long the oil would persist, and how sensitive the shoreline is. The floating oil slicks put the shoreline at particular risk when they eventually come ashore, covering the substrate with oil. The differing substrates between shoreline types vary in their response to oiling, and influence the type of cleanup that will be required to effectively decontaminate the shoreline. In 1995, the US National Oceanic and Atmospheric Administration extended ESI maps to lakes, rivers, and estuary shoreline types. The exposure the shoreline has to wave energy and tides, substrate type, and slope of the shoreline are also taken into account – in addition to biological productivity and sensitivity. The productivity of the shoreline habitat is also taken into account when determining ESI ranking. Mangroves and marshes tend to have higher ESI rankings due to the potentially long-lasting and damaging effects of both the oil contamination and cleanup actions. Impermeable and exposed surfaces with high wave action are ranked lower due to the reflecting waves keeping oil from coming onshore, and the speed at which natural processes will remove the oil.

Biological Resources

Habitats of plants and animals that may be at risk from oil spills are referred to as “elements” and are divided by functional group. Further classification divides each element into species groups with similar life histories and behaviours relative to their vulnerability to oil spills. There are eight element groups: Birds, Reptiles Amphibians, Fish, Invertebrates, Habitats and Plants, Wetlands, and Marine Mammals and Terrestrial Mammals. Element groups are further divided into sub-groups, for example, the ‘marine mammals’ element group is divided into dolphins, manatees, pinnipeds (seals, sea lions & walruses), polar bears, sea otters and whales. Problems taken into consideration when ranking biological resources include the observance of a large number of individuals in a small area, whether special life stages occur ashore (nesting or molting), and whether there are species present that are threatened, endangered or rare.

Human-use Resources

Human use resources are divided into four major classifications; archaeological importance or cultural resource site, high-use recreational areas or shoreline access points, important protected management areas, or resource origins. Some examples include airports, diving sites, popular beach sites, marinas, natural reserves or marine sanctuaries.

Estimating the Volume of a Spill

By observing the thickness of the film of oil and its appearance on the surface of the water, it is possible to estimate the quantity of oil spilled. If the surface area of the spill is also known, the total volume of the oil can be calculated.

	Film thickness			*Quantity spread*	
Appearance	***in***	***mm***	***nm***	***gal/sq mi***	***L/ha***
Barely visible	0.0000015	0.0000380	38	25	0.370
Silvery sheen	0.0000030	0.0000760	76	50	0.730
First trace of colour	0.0000060	0.0001500	150	100	1.500
Bright bands of colour	0.0000120	0.0003000	300	200	2.900
Colours begin to dull	0.00004	0.0010000	1000	666	9.700
Colours are much darker	0.0000800	0.0020000	2000	1332	19.500

Oil spill model systems are used by industry and government to assist in planning and emergency decision making. Of critical importance for the skill of the oil spill model prediction is the adequate description of the wind and current fields. There is a worldwide oil spill modelling (WOSM) program.

Tracking the scope of an oil spill may also involve verifying that hydrocarbons collected during an ongoing spill are derived from the active spill or some other source. This can involve sophisticated analytical chemistry focused on finger printing an oil source based on the complex mixture of substances present. Largely, these will be various hydrocarbons, among the most useful being polyaromatic hydrocarbons. In addition, both oxygen and nitrogen heterocyclic hydrocarbons, such as parent and alkyl homologues of carbazole, quinoline, and pyridine, are present in many crude oils. As a result, these compounds have great potential to supplement the existing suite of hydrocarbons targets to fine tune source tracking of petroleum spills. Such analysis can also be used to follow weathering and degradation of crude spills.

Environmental Impact of Petroleum

The environmental impact of petroleum is often negative because it is toxic to almost all forms of life. The possibility of climate change exists. Petroleum, commonly referred to as oil, is closely linked to virtually all aspects of present society, especially for transportation and heating for both homes and for commercial activities.

Issues

Toxicity: Crude oil is a mixture of many different kinds of organic compounds, many of which are highly toxic and cancer causing (carcinogenic). Oil is "acutely lethal" to fish, that is it kills fish quickly, at a concentration of 4000 parts per million (ppm) (0.4%). "It only takes one quart of motor oil to make 250,000 gallons of ocean water toxic to wildlife." This is the equivalent of a concentration of 1 ppm. Crude oil and petroleum distillates cause birth defects.

Benzene is present in both crude oil and gasoline and is known to cause leukemia in humans. The compound is also known to lower the white blood cell count in humans, which would leave people exposed to it more susceptible to infections. "Studies have linked benzene exposure in the mere parts per billion (ppb) range to terminal leukemia, Hodgkins lymphoma, and other blood and immune system diseases within 5-15 years of exposure."

Exhaust

When oil or petroleum distillates are burned, usually the combustion is not complete. This means that incompletely burned compounds are created in addition to just water and carbon dioxide. The other compounds are often toxic to life. Examples are carbon monoxide and methanol. Also, fine particulates of soot blacken humans' and other animals' lungs and cause heart problems or death. Soot is cancer causing (carcinogenic).

Acid rain

High temperatures created by the combustion of petroleum cause nitrogen gas in the surrounding air to oxidize, creating nitrous oxides. Nitrous oxides, along with sulfur dioxide from the sulfur in the oil, combine with water in the atmosphere to create acid rain. Acid rain causes many problems such as dead trees and acidified lakes with dead fish. Coral reefs in the world's oceans are killed by acidic water caused by acid rain. Acid rain leads to increased corrosion of machinery and structures (large amounts of capital), and to the slow destruction of important archaeological structures such as the marble ruins in Rome and Greece.

Climate Change

Humans burning large amounts of petroleum create large amounts of CO_2 (carbon dioxide) gas that traps heat in the earth's atmosphere.

Also certain organic compounds, such as methane released from petroleum drilling or from the petroleum itself, trap heat several times more efficiently than CO_2. Soot blocks the sun from reaching the earth and could cause cooling of the earth's atmosphere. Oil spills are the release of liquid petroleum hydrocarbons into the environment due to human activity, and are a form of pollution. The term often refers to marine oil spills, where oil is released into the ocean or coastal waters. Oil spills include releases of crude oil from tankers, offshore platforms, drilling rigs and wells, as well as spills of refined petroleum products (such as gasoline, diesel) and their by-products, as well as heavier fuels used by large ships such as bunker fuel, or the spill of any oily refuse or waste oil. Spills may take years or even decades to clean up, and their total environmental impacts are not completely understood.

Oil Shale

Winning of petroleum from oil shale creates pollution problems

Volatile Organic Compounds

Volatile organic compounds (VOCs) are gases or vapours emitted by various solids and liquids, many of which have short- and long-term adverse effects on human health and the environment. VOCs from petroleum are toxic and foul the air, and some like benzene are extremely toxic, carcinogenic and cause DNA damage. Benzene often makes up about 1% of crude oil and gasoline. Benzene is present in automobile exhaust.

Waste Oil

Waste oil is used oil containing breakdown products and impurities from use. Some examples of waste oil are used oils such as hydraulic oil, transmission oil, brake fluids, motor oil, crankcase oil, gear box oil and synthetic oil. Many of the same problems associated with natural petroleum exist with waste oil. When waste oil from vehicles drips out engines over streets and roads, the oil travels into the water table bringing with it such toxins as benzene. This poisons both soil and drinking water. Runoff from storms carries waste oil into rivers and oceans, poisoning them as well.

Mitigation

Conservation/Phasing Out:

- Creating laws to completely phase out the use of petroleum (Sweden's 15 year plan)

- Making use of petroleum more efficiently via better technology.

Substitution of other Energy Sources

- Using "cleaner" energy sources such as natural gas and biodiesel, especially in critical areas like cities where there are people.

Use of Biomass Instead of Petroleum

- It is said that anything that can be made from oil can be made from cellulose, that is fibrous plant material such as from hemp.
- Plastics can be created from cellulose instead of from oil.
- Lubricants like motor oil and grease can be made from plants and animal fat.

Safety Measures

- Decreasing the risk of spills
- False floors at gasoline stations to catch gasoline and oil drips from making it into the water table

Oil Pollution Act of 1990

The Oil Pollution Act (101 H.R.1465, P.L. 101-380) was passed by the 101st United States Congress, and signed by President George H. W. Bush, to mitigate and prevent civil liability for future oil spills off the coast of the United States.

The law stated that companies must have a "plan to prevent spills that may occur" and have a "detailed containment and cleanup plan" for oil spills. The law also includes a clause that prohibits any vessel that, after March 22, 1989, has caused an oil spill of more than one million U.S. gallons (3,800 m^3) in any marine area, from operating in Prince William Sound.

History

The bill was introduced to the House by Walter B. Jones, Sr., a Democratic congressman from North Carolina's 1st congressional district, along with 79 cosponsors following the 1989 Exxon Valdez oil spill, which at the time was the largest oil spill in U.S. history. It enjoyed widespread support, passing the House 375-5 and the Senate by voice vote before conference, and unanimously in both chambers after conference. The U.S. Constitution, as interpreted in *Gibbons v. Ogden* (1824), gives Congress the sole authority to regulate navigable waters.

In April 1998, Exxon argued in a legal action against the federal government that the Exxon Valdez should be allowed back into Alaskan waters. Exxon claimed the OPA was effectively a bill of attainder, a regulation that was unfairly directed at Exxon alone. In 2002, the 9th Circuit Court of Appeals ruled against Exxon. As of 2002, OPA had prevented 18 ships from entering Prince William Sound.

Enforcement

Following the 2010 Deepwater Horizon oil spill in the Gulf of Mexico, numerous U.S. Senators attempted to pass a bill to raise the $75 million cap limit to $10 billion, retroactive to before the spill occurred. Senators of both Republican Party and Democratic Party blocked efforts for new legislation on multiple occasions, arguing that the new law could have unintended consequences. Democratic Party senator Mary Landrieu of Louisiana was quoted in saying “We want to be careful before we change any of these laws that we don’t jeopardize the operations of an ongoing industry, because there are 4,000 other wells in the Gulf that have to go on.” This statute limits BP’s monetary damages to $75 million for losses to private parties, although it still remains liable for all cleanup costs under the law.

National Oil and Hazardous Substances Pollution Contingency Plan

The National Oil and Hazardous Substances Pollution Contingency Plan, more commonly called the National Contingency Plan or NCP, is the United States federal government’s blueprint for responding to oil spills and hazardous substance releases. It documents national response capability and is intended to promote overall coordination among the hierarchy of responders and contingency plans.

The first National Contingency Plan was developed and published in 1968 in response to a massive oil spill from the oil tanker Torrey Canyon off the coast of England that occurred a year earlier. More than 37 million gallons of crude oil spilled into the water and caused massive environmental damage. To avoid the problems faced by response officials involved in the incident, U.S. officials developed a coordinated approach to cope with potential spills in U.S. waters. The

1968 plan provided the first comprehensive system of accident reporting, spill containment, and cleanup. It also established a response headquarters, a national reaction team, and regional reaction teams (precursors to the current National Response Team and Regional Response Teams).

Congress has broadened the scope of the National Contingency Plan over the years. As required by the Clean Water Act of 1972, the NCP was revised the following year to include a framework for responding to hazardous substance spills as well as oil discharges.

Following the passage of Superfund legislation in 1980, the NCP was broadened to cover releases at hazardous waste sites requiring emergency removal actions. Over the years, additional revisions have been made to the NCP to keep pace with the enactment of legislation. The latest revisions to the NCP were finalized in 1994 to reflect the oil spill provisions of the Oil Pollution Act of 1990.

Under the National Contingency Plan, federal agencies should: (1) Plan for emergencies and develop procedures for addressing oil discharges and releases of hazardous substances, pollutants, or contaminants; (2) Coordinate their planning, preparedness, and response activities with one another; (3) Coordinate their planning, preparedness, and response activities with affected states, local governments, and private entities; and (4) Make available those facilities or resources that may be useful in a response situation, consistent with agency authorities and capabilities .

Once a response has been triggered, the USCG or USEPA "is authorized to initiate and, in the case of a discharge posing a substantial threat to public health or welfare of the United States is required to initiate and direct, appropriate response activities when the Administrator or Secretary determines that any oil or CWA hazardous substance is discharged or there is a substantial threat of such discharge from any vessel or offshore or onshore facility into or on the navigable waters of the United States, on the adjoining shorelines to the navigable waters, into or on the waters of the exclusive economic zone, or that may affect natural resources belonging to, appertaining to, or under exclusive management authority of the United States" .

The federal On-Scene Coordinator (OSC) "directs response efforts and coordinates all other efforts at the scene of a discharge or release"

Major Revisions to 40 CFR Part 300

- 1968 Initial plan
- 1973 Added provisions for hazardous substances
- 1980 (circa) revised to reflect CERCLA provisions
- 1990 Revised and reorganized to reflect SARA provisions
- 1994 Revised to reflect OPA revisions.

Role in BP oil Spill

The plan places responsibility for command and control in managing serious disaster response with the U.S. Federal government and not a private company like British Petroleum according to a 2010 article in *Rolling Stone (magazine)* about the BP Gulf oil spill.

Key Provisions of National Contingency Plan

Establishes the National Response Team and its roles and responsibilities in the National Response system, including planning and coordinating responses to major discharges of oil or hazardous waste, providing guidance to Regional Response Teams, coordinating a national program of preparedness planning and response, and facilitating research to improve response activities. EPA serves as the lead agency within the National Response Team (NRT). §300.115 Establishes the Regional Response Teams and their roles and responsibilities in the National Response System, including, coordinating preparedness, planning, and response at the regional level. The RRT consists of a standing team made up of representatives of each federal agency that is a member of the NRT, as well as state and local government representatives, and also an incident-specific team made up of members of the standing team that is activated for a response. The RRT also provides oversight and consistency review for area plans within a given region.

- 300.120 Establishes general responsibilities of federal On-Scene Coordinators.
- 300.125(a) Requires notification of any discharge or release to the National Response Centre through a toll-free telephone number. The National Response Center (NRC) acts as the central clearinghouse for all pollution incident reporting.
- 300.135(a) Authorizes the predesignated On-Scene Coordinator to direct all federal, state, and private response activities at the site of a discharge.

- 300.135(d) Establishes the unified command structure for managing responses to discharges through coordinated personnel and resources of the federal government, the state government, and the responsible party.
- 300.165 Requires the On-Scene Coordinator to submit to the RRT or NRT a report on all removal actions taken at a site.
- 300.170 Identifies the responsibilities for federal agencies that may be called upon during response planning and implementation to provide assistance in their respective areas of expertise consistent with the agencies' capabilities and authorities.
- 300.175 Lists the federal agencies that have duties associated with responding to releases.
- 300.210 Defines the objectives, authority, and scope of Federal Contingency Plans, including the National Contingency Plan (NCP), Regional Contingency Plans (RCPs), and Area Contingency Plans (ACPs).

Oil Removals §300.317 Establishes national priorities for responding to a release.

§300.320 Establishes the general pattern of response to be executed by the On-Scene Coordinator (OSC), including determination of threat, classification of the size and type of the release, notification of the RRT and the NRC, and supervision of thorough removal actions.

§300.322 Authorizes the OSC to determine whether a release poses a substantial threat to the public health or welfare of the United States based on several factors, including the size and character of the discharge and its proximity to human populations and sensitive environments. In such cases, the OSC is authorized to direct all federal, state, or private response and recovery actions. The OSC may enlist the support of other federal agencies or special teams.

§300.323 Provides special consideration to discharges which have been classified as a spill of national significance. In such cases, senior federal officials direct nationally-coordinated response efforts.

§300.324 Requires the OSC to notify the National Strike Force Coordination Center (NSFCC) in the event of a worst case discharges, defined as the largest foreseeable discharge in adverse weather conditions. The NSFCC coordinates the acquisition of needed response personnel and equipment. The OSC also must require implementation of the worst case portion of the tank vessel and Facility Response Plans and the Area Contingency Plan.

§300.355 Provides funding for responses to oil releases under the Oil Spill Liability Trust Fund, provided certain criteria are met. The responsible party is liable for federal removal costs and damages as detailed in section 1002 of the Oil Pollution Act (OPA). Federal agencies assisting in a response action may be reimbursed. Several other federal agencies may provide financial support for removal actions.

Subpart J Establishes the NCP Product Schedule, which contains dispersants and other chemical or biological products that may be used in carrying out the NCP. Authorization for the use of these products is conducted by Regional Response Teams and Area Committees, or by the OSC in consultation with EPA representatives.

Hazardous Substance Removals §300.415(b) Authorizes the lead agency to initiate appropriate removal action in the event of a hazardous substance release. Decisions of action will be based on threats to human or animal populations, contamination of drinking water supplies or sensitive ecosystems, high levels of hazardous substances in soils, weather conditions that may cause migration or release of hazardous substances, the threat of fire or explosion, or other significant factors effecting the health or welfare or the public or the environment.

§300.415(c) Authorizes the OSC to direct appropriate actions to mitigate or remove the release of hazardous substances.

Oil Spill Governance in the United States

The U.S. is an active player in the oil industries. US oil requirements exceed its production rate hence it imports oil from other countries. Oil companies in the US engage in oil exploration both offshore and onshore. A natural consequence of these activities is oil spillage which results in the release of crude oil into the environment from wells, drilling rigs, offshore platforms and oil carrying vessels (e.g. Tankers).

During the past three decades, oil consumption and importation in the U.S has been on a steady rising trend while the number of oil spill incidents and volume of oil spilled have been on the decline. The April 2010 deep horizon Gulf of Mexico oil spill has been the only anomaly in the declining trend. The declining trend can be attributed to changes in oil spill legislation after the occurrence of the Exxon

Valdez oil spill which highlighted weak governance in the area of oil spills in the U.S. The subsequent comprehensive reform of oil spill legislation by congress in response to the Exxon Valdez spill produced the Oil Pollution Act of 1990 which has significantly improved governance in this area.

A number of federal authorities are responsible for governing oil spills in the United States. These main actors in the governing system are a combination of state, federal and international authorities. They are collectively responsible for creating and implementing legislation to prevent oil spills and handling the aftermath decisions and procedures that follow.

Timeline of Key Events

The governance framework for oil spills in the United States prior to the Oil Pollution Act (OPA) of 1990 lacked proper consolidation and proved inadequate in preventing and responding to oil spillages. A timeline produced by the World Resources Institute (WRI) Highlight some key events in oil drilling governance and regulatory system which ultimately govern oil spills in the United States

- 1978: Congress passed amendments to the Outer Continental Shelf Lands Act, Charging the Secretary of interior with overseeing the developments of offshore oil reserves, according to a new tiered management structure.
- 1978: The White House council on Environmental Quality issued regulations regarding federal agencies mandatory implementation of the National Environmental Policy Act of 1969
- 1985: The council on Environmental Quality approved the MMS departmental manual, which established the agency's policy for NEPA implementation and issuing categorical exclusions.
- 1986: The Council on Environmental Quality revoked its 1978 requirement that agencies include "worst-case analyses" in Environmental Impact Statements for actions with unknown or questionable risks.
- 1989: March 24, the Exxon Valdez oil tanker struck a reef off the coast of Alaska, spilling millions of gallons of crude oil into the Prince William Sound
- 1990: President George H.W Bush expanded congressional moratoria on offshore drilling.

- 1990: The Oil Pollution Act of 1990 was passed by congress as a result of the Exxon Valdez oil spill.
- 1994: June 30, MMS adopted a voluntary "safety systems management model" developed by the American Petroleum Institute.
- 2000: March 3, MMS issued a Safety Alert urging offshore lease holders to install a backup mechanism for activating a rig's blowout preventer (BOP) in the event of a blowout to prevent oil spillage.
- 2003: MMS reversed the Safety Alert making having a backup system for remotely activating a rig's BOP a non requirement.

Although the OPA provides a framework for oil spillage governance recent events resulting in the BP deep water oil spill in the Gulf of Mexico suggest more can and should have been done in this area. Evidence of this fact is present in the recommendations and investigations present in the January 2011 report by the recently established National Oil Spill Commission. A further sign of lack of proper governance is implicit in the action of the federal government to rename and reform the Minerals Management Service (MMS) into the now Bureau of Ocean Energy Management, Regulation and Enforcement.

Governance Framework

Legislation Governing Oil Spills in the U.S.: The 1989 Exxon Valdez oil spill served as a major focal point in oil spill governance and reform. Prior to the Exxon Valdez incident a number of laws were in place to prevent and respond to oil spills. These include:

- The National Oil and Hazardous Substances Pollution Contingency Plan (NCP) 1968: The NCP established the response system the federal government was to follow in the event of oil spills and release of hazardous materials into the environment. The NCP was a response by U.S policy makers to the Torrey Canyon oil tanker spill off the coast of England. It has since been amended by the Clean Water Act (1972), the Oil Pollution Act (1990) and the Comprehensive Environmental Response, Competition and Liability Act (CERCLA) 1980. The Oil Pollution Act increased the role and dimensions of the NCP by establishing a more robust planning and response system to improve response and prevent spills in marine environments.

- The Clean Water Act (1972): The clean water act (CWA) was the most extensive legislation which considered oil spills prior the Exxon Valdez spill. The CWA made provisions for post spill reporting, response and liability by the responsible party.
- The Trans-Alaska Pipeline Authorization Act (1973): Major oil transportation via pipelines is goes through the Trans-Alaskan route. Spills from pipelines along this route although inland, could migrate into coastal waters via inland rivers. Hence the act was established to cover oil spills and liability relating to the Trans-Alaska Pipeline System (TAPS).
- The Deep Port Act (1974): This was the major statute for deep water spill incidents. It addressed oil spills clean up and liability at deepwater oil ports.
- The Outer Continent Shelf Lands Act Amendments (1978): This act addressed oil spills, clean up and liability structure for oil extraction facilities in federal offshore waters.
- The Hazardous Liquid Pipeline Act of (1979): This act granted the Department of Transport (DOT) authority to govern oil spills from pipelines.
- The Pipeline Safety Improvement Act of (2006): This act was established to improve pipeline safety and security practices. This act also reaffirmed the role of the federal office of pipeline safety relevant governing body in terms of pipeline spills under the DOT.

The above legislations were the principle governing instruments of oil spills incidents prior to the Exxon Valdez oil spill. Several attempts by congress to establish more encompassing and elaborate oil pollution laws were hindered by belligerent issues which mostly produced stalemates. One such conflict was federal law limiting a state's ability to enforce requirements and liability for parties responsible for causing oil spills. The focus of the debates was mostly centred on party liability. An example of the type of debates is illustrated in the question "who is liable in the case of a vessel spill?" .The cargo owner or the ship operator/ owner? Another significant issue was the interaction of domestic legislation and international measures. In 1980's, international agreements being considered would take over oil spills federal and state laws adding further complexity to party liability.

However after the Exxon Valdez incident, the short comings of the patchy framework for oil spill governance was apparent and growing pressure placed on lawmakers resulted in the establishment of the more comprehensive Oil Pollution Act of 1990.

The Oil Pollution Act (1990): The OPA is the primary legislation that governs oil spills in the U.S. The establishment of the OPA substantiated the federal government's role in responding to oil spill cleanups. The OPA made amendments to the already existing CWA to provide 3 options to the delegated authorities through the president. The options include conducting immediate cleanup by federal authorities, to monitor the response of the responsible party or commandeer the cleanup activities of the responsible party. Hence giving the federal government the authority to determine the level of clean up required.

A second significant amendment the OPA makes to the CWA is that it makes it a mandatory requirement for U.S tank vessels, onshore and offshore facilities to establish and submit oil spill response plans to the corresponding federal authority. The OPA requires new vessels operating and transporting oil in U.S waters to have double hulls. Not all vessels are of the same size and for the same purpose hence the OPA makes provision for exempting certain vessels from having double hulls. However, by 2015 it is a requirement by law that all oil carrying vessels operating in the U.S must have double hulls.

Federal Agencies Responsibility

The legal framework for enforcing oil spill governance in the U.S involves a number of federal authorities. The responsibilities of the agency is split into two categories: (1) Oil spill anticipation and prevention and (2) oil spill response and cleanup

Response and Clean-up

The single most important authority governing oil spills in the United States is the Federal government. The federal government has jurisdiction over oil responses and oil spills that occur in state and federal navigable waters alike.

The location of the Oil spill determines which authority responds. Oil spills that occur in coastal waters are the responsibility of the United States Coast Guard (USCG) while the Environmental Protection Agency covers inland oil spills. It is required by US federal law that any discharge of oil that creates a film or sheen on the water surface be reported to the National Response Centre. The National Response Centre then dissipates information to the USCG who acts as the federal onshore coordinator.

The USCG are the primary federal response authority in coastal water hence having the overall power to ensure the effective clean up of oil spills and lead actions that prevent further discharge from the spill source . The resulting activities that follow up an oil spill involving federal, state and private parties are co-ordinated by the USCG. The National Oceanic and Atmospheric Administration, Office of Response and Restoration (NOAA) works closely with the USCG in providing assistance in technical areas such as Consideration of Alternatives, Oil displacement tracking and risk assessments.

Anticipation and Prevention

Different authorities are responsible for governing oil spill prevention and anticipation depending on the potential sources of the oil spills. A number of executive orders (EOs) and memoranda of understanding (MOU) have established the authorities and agencies responsible for various classes of potential oil spills.

Table 1: *Federal Agency Jurisdiction for oil spill anticipation and prevention duties, by potential sources.*

Potential Source of Oil Spill	*Responsible Agency*
Onshore, non-transportation facilities	EPA
Onshore, transportation facilities	USCG and Department of Transportation (DOT)
Deep water ports	USCG and DOT
Offshore facilities (oil and gas extraction)	Mineral Management Service (MMS) within the Department of Interior
Offshore pipelines directly associated with oil extraction activities (i.e. production lines)	MMS
Offshore pipelines not directly associated with oil extraction activities (i.e. transmission lines)	Office of pipeline Safety (OPS) within the DOT
Inland Pipelines	OPS

The preventive measures taken by relevant authorities include assessment of facilities to ensure the required standards set out by legislation are met e.g. Double hulls in new vessels and secondary containment in oil holding facilities. Anticipatory duties involve managing the response plans of vessels and facilities to oil spills at various levels: state, regional and national. This ensures proper training of personnel on vessels and facilities to carry out their outlined response plans is also a key duty. The OPA requires relevant agencies to conduct examination to test the anticipative capacities of the parties involved.

International Oil Spill Governance in the U.S.

International conventions have played an important role in creating external networks for governing oil spills in the U.S. international treaties when signed by the U.S. are on the same level as federal law, hence signatory parties must implement domestic legislation to reflect the agreement. This mechanism helps to engineer a number of federal laws governing oil spills such as the intervention on the High Sea Act of 1974. International conventions serve a key role in developing standards for oil carrying vessels from different nations into the U.S. The two major conventions which have contributed to oil spill governance are the 1969 International Convention Relating to Intervention on the High Seas in Cases of Oil Pollution Casualties (the Intervention Convention) and the International Convention for the Prevention of Pollution from Ships (MARPOL 73/78).

The most important international organisations that contribute to oil spill governance in the U.S include the International Maritime Organisation (IMO), International Spill Control Organisation (ISCO) and the Association of Petroleum Industry Managers (APICOM).

Governance Approach to U.S. oil Spills

The oil spill framework in the U.S employs a multilateral system of governance where the federal authorities, NGO's and private parties are all actively involved in response and clean up procedures. Although the U.S. government plays an important role in regulating oil spills, it however does not entirely command and control every aspect of the process.

Due to the complex nature of oil and gas operations and limited technical expertise by the government in such industries, industry standard setting and self-policing play a significant role in the governing process of oil spills. This has inevitably led to the network governance approach used to govern oil spills. The insufficient expertise and specialist technical knowledge in the public sector makes it difficult for the government to rely entirely on its personnel hence resulting in public-private partnership known as a type II partnership. This style of new public management to oil spill governance is common in other aspects of environmental governance (e.g. climate change) in the U.S and differs from the typical bureaucratic role of enforcement the federal government usually play.

The governance process of oil spills relies on input from national, private and international institutions. To effectively implement response, prevention and clean up procedures a number of public authorities were given responsibility under different jurisdictions. Since the 1970s the number of institutions governing oil spills has increased, indicating a steady shift from state led government approach to a broader and multi layered governance process. The implementation of international oil spill treaties into domestic legislation discussed earlier provides further evidence of the shift to a more governance based approach.

Ohmsett

Ohmsett is the National Oil Spill Response Test Facility, located in Leonardo, New Jersey. The name Ohmsett is an acronym for "Oil and Hazardous Materials Simulated Environmental Test Tank".

This is the only facility of its kind where full-scale oil spill response equipment testing, research, and training can be conducted in a marine environment with oil under controlled environmental conditions. Variables such as waves, temperature, and oil types are able to be controlled. A benefit of this facility is that it provides an environmentally safe place to conduct objective testing and to develop devices and techniques for the control of oil and hazardous material spills.

The mission of Ohmsett is to strengthen awareness of oil spill pollution prevention and response methods, while at the same time remaining committed to the well being of its customers, employees, and associates.

The facility, located an hour south of New York City, in Leonardo, New Jersey, is maintained and operated by the Minerals Management Service (MMS), a bureau in the U.S. Department of the Interior (which is a Federal agency) that manages the nation's natural gas, oil and other mineral resources on the outer continental shelf (OCS). They do this through a contract with MAR, Incorporated.

History

The Ohmsett facility was originally built by the U.S. Environmental Protection Agency (EPA) in 1974 and was operated by that same agency until 1987. At that time, it was known as the Oil and Hazardous Materials Simulated Environmental Test Tank, or OHMSETT. It is now just referred to by the acronym. In 1989, Ohmsett, as it was known then, was closed and responsibility for the facility was transferred to the U.S. Navy (USN). This was done because the facility is located on the Naval Weapons Station Earle in Leonardo, NJ.

Title VII of the Ocean Pollution Act of 1990 (OPA 90) gave MMS the lead responsibility for reactivation of Ohmsett in the aftermath of the Exxon Valdez oil spill in Alaska. The MMS was also charged with the continuing operation and maintenance of the facility as a national test facility. The MMS refurbished Ohmsett beginning in 1990 and reopened it for testing in 1992. Costs for the yearly operation and maintenance of Ohmsett are covered by the Oil Spill Liability Trust Fund (OSTLF). The OSTLF derives its funds from a tax on companies that produce or transport oil. Because of this, no appropriated taxpayer dollars are used to support this unique oil spill response technology testing, training, and research facility.

2

Petrochemical

Petrochemicals are chemical products derived from petroleum. Some chemical compounds made from petroleum are also obtained from other fossil fuels, such as coal or natural gas, or renewable sources such as corn or sugar cane. This article focuses on organic compounds that are not burned as fuel.

Two petrochemical classes are olefins including ethylene and propylene, and aromatics including benzene, toluene and xylene isomers. Oil refineries produce olefins and aromatics by fluid catalytic cracking of petroleum fractions. Chemical plants produce olefins by steam cracking of natural gas liquids like ethane and propane. Aromatics are produced by catalytic reforming of naphtha. Olefins and aromatics are the building blocks for a wide range of materials such as solvents, detergents, and adhesives. Olefins are the basis for polymers and oligomers used in plastics, resins, fibres, elastomers, lubricants, and gels.

Global ethylene and propylene production are ~110 million tonnes and ~65 million tonnes per annum respectively. Aromatics production is ~70 million tonnes. The largest petrochemical industries are located in the USA and Western Europe; however, major growth in new production capacity is in the Middle East and Asia. There is substantial inter-regional petrochemical trade.

Primary petrochemicals are divided into three groups depending on their chemical structure:

- Olefins includes ethylene, propylene, and butadiene. Ethylene and propylene are important sources of industrial chemicals and plastics products. Butadiene is used in making synthetic rubber.

- Aromatics includes benzene, toluene, and xylenes. Benzene is a raw material for dyes and synthetic detergents, and benzene and toluene for isocyanates MDI and TDI used in making polyurethanes. Manufacturers use xylenes to produce plastics and synthetic fibers.
- Synthesis gas is a mixture of carbon monoxide and hydrogen used to make ammonia and methanol. Ammonia is used to make the fertilizer urea and methanol is used as a solvent and chemical intermediate.

The prefix "petro-" is an arbitrary abbreviation of the word "petroleum"; since "petro-" is Ancient Greek for "rock" and "oleum" means "oil". Therefore, the etymologically correct term would be "oleochemicals". However, the term oleochemical is used to describe chemicals derived from plant and animal fats.

Significant Petrochemicals and their Derivatives

The following is a partial list of the major commercial petrochemicals and their derivatives:

- ethylene - the simplest olefin; used as a chemical feedstock and ripening hormone
  - polyethylene - polymerized ethylene
  - ethanol - via ethylene hydration (chemical reaction adding water) of ethylene
  - ethylene oxide - via ethylene oxidation
    - ethylene glycol - via ethylene oxide hydration
    - engine coolant - ethylene glycol, water and inhibitor mixture
    - polyesters - any of several polymers with ester linkages in the backbone chain
    - glycol ethers - via glycol condensation
    - ethoxylates
  - vinyl acetate
  - 1,2-dichloroethane
    - trichloroethylene

— tetrachloroethylene - also called perchloroethylene; used as a dry cleaning solvent and degreaser

— vinyl chloride - monomer for polyvinyl chloride

— polyvinyl chloride (PVC) - type of plastic used for piping, tubing, other things

- propylene - used as a monomer and a chemical feedstock
  - o isopropyl alcohol - 2-propanol; often used as a solvent or rubbing alcohol
  - o acrylonitrile - useful as a monomer in forming Orlon, ABS
  - o polypropylene - polymerized propylene
  - o propylene oxide
    - — polyol - used in the production of polyurethanes
    - — propylene glycol - used in engine coolant and aircraft deicer fluid
    - — glycol ethers - from condensation of glycols
  - o acrylic acid
    - — acrylic polymers
  - o allyl chloride -
    - — epichlorohydrin - chloro-oxirane; used in epoxy resin formation
    - — epoxy resins - a type of polymerizing glue from bisphenol A, epichlorohydrin, and some amine
- C4 hydrocarbons - a mixture consisting of butanes, butylenes and butadienes
  - o isomers of butylene - useful as monomers or co-monomers
    - — isobutylene - feed for making methyl *tert*-butyl ether (MTBE) or monomer for copolymerization with a low percentage of isoprene to make butyl rubber
  - o 1,3-butadiene (or buta-1,3-diene) - a diene often used as a monomer or co-monomer for polymerization to elastomers such as polybutadiene, styrene-butadiene rubber, or a plastic such as acrylonitrile-butadiene-styrene (ABS)
    - — synthetic rubbers - synthetic elastomers made of any one or more of several petrochemical (usually) monomers such as 1,3-butadiene, styrene, isobutylene,

isoprene, chloroprene; elastomeric polymers are often made with a high percentage of conjugated diene monomers such as 1,3-butadiene, isoprene, or chloroprene

- higher olefins
  - o polyolefins such poly-alpha-olefins which are used as lubricants
  - o alpha-olefins - used as monomers, co-monomers, and other chemical precursors. For example, a small amount of 1-hexene can be copolymerized with ethylene into a more flexible form of polyethylene.
  - o other higher olefins
  - o detergent alcohols
- benzene - the simplest aromatic hydrocarbon
  - o ethylbenzene - made from benzene and ethylene
    - — styrene made by dehydrogenation of ethylbenzene; used as a monomer
    - — polystyrenes - polymers with styrene as a monomer
  - o cumene - isopropylbenzene; a feedstock in the cumene process
    - — phenol - hydroxybenzene; often made by the cumene process
    - — acetone - dimethyl ketone; also often made by the cumene process
    - — bisphenol A - a type of "double" phenol used in polymerization in epoxy resins and making a common type of polycarbonate
    - — epoxy resins - a type of polymerizing glue from bisphenol A, epichlorohydrin, and some amine
    - — polycarbonate - a plastic polymer made from bisphenol A and phosgene (carbonyl dichloride)
    - — solvents - liquids used for dissolving materials; examples often made from petrochemicals include ethanol, isopropyl alcohol, acetone, benzene, toluene, xylenes

- o cyclohexane - a 6-carbon aliphatic cyclic hydrocarbon sometimes used as a non-polar solvent
  - — adipic acid - a 6-carbon dicarboxylic acid which can be a precursor used as a co-monomer together with a diamine to form an alternating copolymer form of nylon.
  - — nylons - types of polyamides, some are alternating copolymers formed from copolymerizing dicarboxylic acid or derivatives with diamines
  - — caprolactam - a 6-carbon cyclic amide
  - — nylons - types of polyamides, some are from polymerizing caprolactam
- o nitrobenzene - can be made by single nitration of benzene
  - — aniline - aminobenzene
    - * methylene diphenyl diisocyanate (MDI) - used as a co-monomer with diols or polyols to form polyurethanes or with di- or polyamines to form polyureas
  - — polyurethanes
- o alkylbenzene - a general type of aromatic hydrocarbon which can be used as a presursor for a sulfonate surfactant (detergent)
  - _ detergents - often include surfactants types such as alkylbenzenesulfonates and nonylphenol ethoxylates
- o chlorobenzene

• toluene - methylbenzene; can be a solvent or precursor for other chemicals
- o benzene
- o toluene diisocyanate (TDI) - used as co-monomers with diols or polyols to form polyurethanes or with di- or polyamines to form polyureas
  - — polyurethanes - a polymer formed from diisocyanates and diols or polyols
- o benzoic acid - carboxybenzene
  - — caprolactam
    - * nylon.

Product

Product(s) are formed during chemical reactions as reagents are consumed. Products have lower energy than the reagents and are produced during the reaction according to the second law of thermodynamics.

The released energy comes from changes in chemical bonds between atoms in reagent molecules and may be given off in the form of heat or light. Products are formed as the chemical reaction progresses toward chemical equilibrium at a certain reaction rate, which depends on the reagents and environmental conditions.

Depending on the relative amounts of the reagents and the equilibrium of the reaction, the terms "reagent" and "product" may also overlap.

Reagent

A reagent is a "substance or compound that is added to a system in order to bring about a chemical reaction or is added to see if a reaction occurs." Although the terms reactant and reagent are often used interchangeably, a reactant is less specifically a "substance that is consumed in the course of a chemical reaction". Solvents and catalysts, although they are involved in the reaction, are usually not referred to as reactants.

In organic chemistry, reagents are compounds or mixtures, usually composed of inorganic or small organic molecules, that are used to affect a transformation on an organic substrate. Examples of organic reagents include the Collins reagent, Fenton's reagent, and Grignard reagent. There are also *analytical reagents* which are used to confirm the presence of another substance. Examples of these are Fehling's reagent, Millon's reagent and Tollens' reagent.

In another use of the term, when purchasing or preparing chemicals, reagent-grade describes chemical substances of sufficient purity for use in chemical analysis, chemical reactions or physical testing. Purity standards for reagents are set by organizations such as ASTM International. For instance, reagent-quality water must have very low levels of impurities like sodium and chloride ions, silica, and bacteria, as well as a very high electrical resistivity.

- mixed xylenes - any of three dimethylbenzene isomers, could be a solvent but more often precursor chemicals

- o *ortho*-xylene - both methyl groups can be oxidized to form (*ortho*-)phthalic acid
  - — phthalic anhydride
- o *para*-xylene - both methyl groups can be oxidized to form terephthalic acid
  - — dimethyl terephthalate - can be copolymerized to form certain polyesters
  - — polyesters - although there can be many types, polyethylene terephthalate is made from petrochemical products and is very widely used.
  - — purified terephthalic acid - often copolymerized to form polyethylene terephthalate
  - — polyesters
- o *meta*-xylene
  - — isophthalic acid
  - — alkyd resins
  - — Polyamide Resins
  - — Unsaturated Polyesters.

Limiting Reagent

In a chemical reaction, the limiting reagent, also known as the "limiting reactant", is the substance which is totally consumed when the chemical reaction is complete. The amount of product formed is limited by this reagent since the reaction cannot proceed further without it. The other reagents may be present in excess of the quantities required to react with the limiting reagent.

The limiting reagent must be identified in order to calculate the percentage yield of a reaction, since the theoretical yield is defined as the amount of product obtained when the limiting reagent reacts completely.

Given the balanced chemical equation which describes the reaction, there are several equivalent ways to identify the limiting reagent and evaluate the excess quantities of other reagents.

Method 1: Comparison of Reactant Amounts

This method is most useful when there are only two reactants. One reactant (A) is chosen, and the balanced chemical equation is used to determine the amount of the other reactant (B) necessary to react with A. If the amount of B actually present exceeds the amount required, then B is in excess and A is the limiting reagent. If the amount of B present is less than required, then B is the limiting reagent.

Example for Two Reactants

Consider the combustion of benzene, represented by the following chemical equation:

$$2C_6H_6 + 15O_2 \rightarrow 12CO_2 + 6H_2O$$

This means that 15 mol molecular oxygen (O_2) is required to react with 2 mol benzene (C_6H_6).

The amount of oxygen required for other quantities of benzene can be calculated using cross-multiplication (the rule of three). For example, if 1.5 mol C_6H_6 is present, 11.25 mol O_2 is required:

$$1.5 \text{ mol} C_6H_6 \times \frac{15 \text{ mol} O_2}{2 \text{ mol} C_6H_6} = 11.25 \text{ mol} O_2$$

If in fact 18 mol O_2 are present, there will be an excess of (18 - 11.25) = 6.75 mol of unreacted oxygen when all the benzene is consumed. Benzene is then the limiting reagent.

This conclusion can be verified by comparing the mole ratio of O_2 and C_6H_6 required by the balanced equation with the mole ratio actually present:

$$\text{Required: } \frac{molO_2}{molC_6H_6} = \frac{15molO_2}{2molC_6H_6} = 7.5molO_2$$

$$\text{Actual: } \frac{molO_2}{molC_6H_6} = \frac{18molO_2}{1.5molC_6H_6} = 12molO_2$$

Since the actual ratio is larger than required, O_2 is the reagent in excess, which confirms that benzene is the limiting reagent.

Method 2: Comparison of Product Amounts Which can be Formed from Each Reactant

In this method the chemical equation is used to calculate the amount of one product which can be formed from each reactant in the amount present. This method can be extended to any number of reactants more easily than the first method.

Example

Which reactant is limiting if 20.0 g of iron (III) oxide (Fe_2O_3) are reacted with 8.00 g aluminium (Al) in the following thermite reaction?

$$Fe_2O_3(s) + 2Al(s) \rightarrow 2Fe(l) + Al_2O_3(s)$$

Since the reactant amounts are given in grams, they must be first converted into moles for comparison with the chemical equation, in order to determine how many moles of Fe can be produced from either reactant.

Moles produced of Fe from reactant Fe_2O_3

$$molFe_2O_3 = \frac{gramsFe_2O_3}{g / molFe_2O_3}$$

$$molFe_2O_3 = \frac{20.0g}{159.7g / mol} = 0.125mol$$

$$molFe = 0.125 \text{ mol} Fe_2O_3 \times \frac{2 \text{ mol} Fe}{1 \text{ mol} Fe_2O_3} = 0.250 \text{ mol} Fe$$

Moles produced of Fe from reactant Al

$$molAl = \frac{gramsAl}{g / molAl}$$

$$molAl = \frac{8.00g}{26.98g / mol} = 0.297mol$$

$$molFe = 0.297 \text{ mol} Al \times \frac{2 \text{ mol} Fe}{2 \text{ mol} Al} = 0.297 \text{ mol} Fe$$

There is enough Al to produce 0.297 mol Fe, but only enough Fe_2O_3 to produce 0.250 mol Fe. This means that the amount of Fe actually produced is limited by the Fe_2O_3 present, which is therefore the limiting reagent.

Shortcut

It can be seen from the example above that the amount of product (Fe) formed from each reagent X (Fe_2O_3 or Al) is proportional to the quantity:

$$\frac{\text{Moles of Reagent X}}{\text{Coefficient of Reagent X}}$$

This suggests a shortcut which works for any number of reagents. Just calculate this formula for each reagent, and the reagent that has the lowest value of this formula is the limiting reagent.

Limiting Factor

A limiting factor or limiting resource is a factor that controls a process, such as organism growth or species population, size, or distribution.

The availability of food, predation pressure, or availability of shelter are examples of factors that could be limiting for an organism. An example of a limiting factor is sunlight in the rainforest, where growth is limited to all plants in the understory unless more light becomes available.

A number of potential factors could influence a biological process, but importantly only one is limiting at any one place and time. This recognition that there is always a *single* limiting factor is vital in ecology; and the concept has parallels in numerous other processes.

Some other limiting factors in biology are water availability, temperature, shelter, or predation.

Abiotic Component

In biology and ecology, abiotic components are non-living chemical and physical factors in the environment which affect ecosystems. Abiotic phenomena underlie all of biology.

Abiotic factors, while generally downplayed, can have enormous impact on evolution. Abiotic components are aspects of geodiversity. They can also be recognised as "abiotic pathogens"

From the viewpoint of biology, abiotic influences may be classified as light or more generally radiation, temperature, water, the chemical surrounding composed of the terrestrial atmospheric gases, as well as soil.

The macroscopic climate often influences each of the above. Not to mention pressure and even sound waves if working with marine, or deep underground, biome.

Those underlying factors affect different plants, animals and fungi to different extents. Some plants are mostly water starved, so humidity plays a larger role in their biology. If there is little or no sunlight then plants may wither and die from not being able to get enough sunlight to do phytosynthesis. Many archaebacteria require very high temperatures, or pressures, or unusual concentrations of chemical substances such as sulfur, because of their specialization into extreme conditions. Certain fungi have evolved to survive mostly at the temperature, the humidity, and stability.

For example, there is a significant difference in access to water as well as humidity between temperate rainforests and deserts. This difference in water access causes a diversity in the types of plants and animals that grow in these areas.

3

Petroleum Geology Design

Burford v. Sun Oil Co.

Burford v. Sun Oil Co., 319 U.S. 315 (1943) was a United States Supreme Court case in which the Court created a new doctrine of abstention.

Facts

The Railroad Commission of Texas granted defendant Burford an order which gave him a right to drill four oil wells on a small plot of land on the East Texas oil field. Plaintiff Sun Oil Co. sued Burford in United States District Court for the Western District of Texas, asserting both federal question jurisdiction and diversity jurisdiction, and alleging that the Commission's order denied them Due Process of law under the Fourteenth Amendment.

The Commission was charged by Texas state law with the administration of oil and gas regulations, including production quotas for each field and well. Quotas were set in conjunction with other oil- and gas-producing states. Part of its duties was regulating the spacing of the individual oil wells. Because of the complex geology of oil fields and multiple division of surface rights, one user's oil drilling activity can affect the production of others, which meant that both public and private interests were at stake. Underground water and gas pressure needed to be monitored in order to maintain steady oil production and prevent waste.

Because the rule governing the spacing of wells over the oil field, known as "Rule 37", generated a large amount of litigation among many competing landowners and potential drillers, the Texas Legislature had developed a special system of judicial review for such cases. Commission orders could be appealed to the state district court in Travis County, Texas, and could then be reviewed by the Texas Court of Civil Appeals, and then to the Texas Supreme Court, which had final authority over questions of state law.

The district court of Travis County was given exclusive jurisdiction over appeals of Commission orders, allowing them to specialize in such matters and promoting cooperation between the court and the Commission.

There already existed conflicting statutory interpretations between the state and federal courts in this matter, so the district court chose to abstain in deference to the state and dismiss the case. The United States Court of Appeals for the Fifth Circuit reversed the District Court's dismissal.

Plurality Opinion

Justice Black, writing for the plurality, reversed the judgment of the Fifth Circuit and upheld the District Court's dismissal of the matter. The central issue to the plaintiff's claim was the "reasonableness" of the Commission's order according to state law, which Black agreed was a different standard than constitutional due process.

Black compared the situation in this case to that in *Railroad Commission v. Pullman Co.*, arguing that this was a matter of interpretation of state law which needed to be handled authoritatively by the state court system, and that state court review was adequate, especially in light of the fact that questions of federal law could still be reviewed by the U.S. Supreme Court via writ of certiorari from the Texas Supreme Court.

Holding

Under the doctrine of *Burford* abstention, a federal court sitting in diversity jurisdiction may abstain from hearing the case where the state courts likely have greater expertise in a particularly complex and unclear area of state law which is of special significance to the state, where there is comprehensive state administrative/regulatory procedure, and where the federal issues cannot be decided without delving into state law.

Concurrence

Justice Douglas, concurring, implied that the Court's ruling did not go far enough. He noted that the opinion of the Court as I read it does not hold or even fairly imply that 'the enforcement of state rights created by state legislation and affecting state policies is limited to the state courts.' Any such holding would result in a drastic inroad on diversity jurisdiction-a limitation which I agree might be desirable but which Congress, not this Court, should make. The holding in these cases, however, goes to no such length.

Dissent

Justice Frankfurter dissented, defending the concept of diversity jurisdiction as a method of providing an impartial forum for disputes between parties from different states, and criticizing the majority for denying the parties their day in federal court. He noted that the question of whether to limit or abolish diversity jurisdiction is a question for Congress, and not the courts. He distinguished *Railroad Commission v. Pullman Co.* on the grounds that in that case, the state law question could have displaced the federal issue entirely, and that the Texas Supreme Court had already defined the terms of the statute and the scope of judicial review thereunder.

Stuart Oil Shale Project

The Stuart Oil Shale Project was an oil shale development project in Australia near Gladstone, Queensland. It was Australia's first major attempt since the 1950s to restart commercial use of oil shale. The project was originally developed by Australian companies Southern Pacific Petroleum NL and Central Pacific Minerals NL (SPP/CPM).

History

In 1997 SPP/CPM signed a joint venture agreement with the Canadian company Suncor Energy to develop the Stuart oil shale deposit. Suncor was designated as the project operator. In April 2001, Suncor left the project and SPP/CPM became the sole shareholder of the project. In February 2002, due the restructuring of SPP/CPM, SPP became the holding company for the group's interests, including the Stuart Oil Shale Project. As SPP had granted fixed and floating charges in favour of Sandco Koala LLC in May 2003, the chargee appointed receivers of SPP on 2 December 2003. In February 2004, the Stuart Oil Shale Project was sold by receivers to the newly formed company Queensland Energy Resources., which announced on 21 July 2004 that the plant would be closed for economic and environmental reasons. Greenpeace, which had protested the project, viewed the closure as a major victory.

On 14 August 2008, Queensland Energy Resources announced that it would replace the Alberta Taciuk Process (ATP) of oil shale processing with Paraho II technology and that it was dismantling the ATP-based plant.

Project Stages

The first stage of the project, which cost A$250–360 million, consisted of an oil-shale mine and pilot retorting plant, which was constructed in 1997-1999. The plant, which was in operation from 1999 to 2004, used the ATP retorting technology being the first application of the ATP technology in the world on oil shale. The plant was designed to process 6,000 tonnes of oil shale per day with oil output of 4,500 barrels (720 m^3). From 2000 to 2004 the pilot plant produced over 1.5 million barrels (240×10^3 m^3) of shale oil. After the closure the facility was dismantled.

The second stage with cost of A$600 million was planned to consist of a single commercial-size module four times larger than the first with total capacity of 19,000 barrels (3,000 m^3) of oil products (naphtha and medium shale oil) daily. Originally it was planned to become operational in 2006. The third planned stage was construction of multiple commercial production units with capacity of up to 200,000 barrels of oil products per day. It was envisaged to come on stream during 2010–2013. The environmental impact assessment of stage 2 was suspended in December 2004. By April 2008, plant equipment at the site was being sold off.

Environmental Issues

The project was heavily criticized by environmentalists. Over 20,000 people and 27 environment, tourism and fishing groups opposed the shale oil plant. Greenpeace claimed that greenhouse emissions from the production of shale oil was nearly four times higher than from the production of conventional oil. In response, SPP promised to reduce greenhouse emissions from production of shale oil to 5% below those of conventional oil by stage 3. Greenpeace also claimed that the Stuart Oil Shale Project was a significant source of highly toxic dioxins and would damage the Great Barrier Reef World Heritage

Area during stage 3. Public health concerns were also mentioned. Local residents claimed dioxins emitted from the plant affected their health and that the odour was unacceptable.

Greenpeace

Greenpeace is a non-governmental environmental organization with offices in over 40 countries and with an international coordinating body in Amsterdam, The Netherlands. Greenpeace states its goal is to "ensure the ability of the Earth to nurture life in all its diversity" and focuses its work on world wide issues such as global warming, deforestation, overfishing, commercial whaling and anti-nuclear issues. Greenpeace uses direct action, lobbying and research to achieve its goals. The global organization does not accept funding from governments, corporations or political parties, relying on more than 2.8 million individual supporters and foundation grants. Greenpeace is a founding member of the INGO Accountability Charter; an international nongovernmental organization that intends to foster accountability and transparency of non-governmental organizations.

Greenpeace evolved from the peace movement and anti-nuclear protests in Vancouver, British Columbia, in the early 1970s. On September 15, 1971, the newly-founded Don't Make a Wave Committee sent a chartered ship, *Phyllis Cormack*, renamed *Greenpeace* for the protest, from Vancouver to oppose United States testing of nuclear devices in Amchitka, Alaska. The Don't Make a Wave Committee subsequently adopted the name Greenpeace.

In a few years, Greenpeace spread to several countries and started to campaign on other environmental issues such as commercial whaling and toxic waste. In the late 1970s, the different regional Greenpeace groups formed Greenpeace International to oversee the goals and operations of the regional organizations globally. Greenpeace received international attention during the 80s when the French intelligence agency bombed the *Rainbow Warrior* in Auckland's Waitemata Harbour, one of the most well-known vessels operated by Greenpeace, killing one individual. In the following years, Greenpeace evolved into one of the largest environmental organizations in the world.

Greenpeace is known for its direct actions and has been described as the most visible environmental organization in the world. Greenpeace has raised environmental issues to public knowledge, influenced both the private and the public sector. Greenpeace has also been a source of controversy; its motives and methods have received criticism and the organization's direct actions have sparked legal actions against Greenpeace activists.

History

Origins: In the late 1960s, the U.S. had plans for an underground nuclear weapon test in the tectonically unstable island of Amchitka in Alaska. Because of the 1964 Alaska earthquake, the plans raised some concerns of the test triggering earthquakes and causing a tsunami. Anti-nuclear activists protested against the test on the border of the U.S. and Canada with signs reading "Don't Make A Wave. It's Your Fault If Our Fault Goes". The protests did not stop the U.S. from detonating the bomb.

While no earthquake or tsunami followed the test, the opposition grew when the U.S. announced they would detonate a bomb five times more powerful than the first one. Among the opposers were Jim Bohlen, a veteran who had served the U.S. Navy and Irving and Dorothy Stowe, a Jewish couple, who had recently become Quakers. As members of the Sierra Club Canada, they were frustrated by the lack of action by the organization.

From Irving Stowe, Jim Bohlen learned of a form of passive resistance, "bearing witness", where objectionable activity is protested simply by mere presence. Jim Bohlen's wife Marie came up with the idea to sail to Amchitka, inspired by the anti-nuclear voyages of Albert Bigelow in 1958. The idea ended up in the press and was linked to The Sierra Club. The Sierra Club did not like this connection and in 1970 The Don't Make a Wave Committee was established for the protest. Early meetings were held in the Shaughnessy home of Robert and Bobbi Hunter. Subsequently the Stowe home at 2775 Courtenay St. became the HQ. The first office was opened in a back-room, storefront on Cypress and Bwy SE corner in Kitsilano, (Vancouver) before moving to West 4th at Maple.

There is some debate as to who are the actual founders of The Don't Make a Wave Committee. Researcher Vanessa Timmer has referred the early members as "an unlikely group of loosely organized protestors". According to the current Greenpeace web page, the founders were Dorothy and Irving Stowe, Marie and Jim Bohlen, Ben and Dorothy Metcalfe, and Robert Hunter. The book *The Greenpeace Story* states that the founders were Irving Stowe, Jim Bohlen and Paul

Cote, a law student and peace activist. An interview with Dorothy Stowe, Dorothy Metcalfe, Jim Bohlen and Robert Hunter identifies the founders as Paul Cote, Irving and Dorothy Stowe and Jim and Marie Bohlen. Paul Watson, who also participated in the anti-nuclear protests, maintains that he also was one of the founders. Another early member, Patrick Moore also has stated that he was one of the founders. Greenpeace used to list Moore among "founders and first members" of The Don't Make a Wave Committee but has later stated that while Moore was a significant early member, he was not a founder. According to Moore's own letter he applied to the already existing organization in March 1971. Irving Stowe arranged a benefit concert (supported by Joan Baez) that took place on October 16, 1970 at the Pacific Coliseum in Vancouver. The concert created the financial basis for the first Greenpeace campaign. Amchitka, the 1970 concert that launched Greenpeace has been published by Greenpeace in November 2009 on CD and is also available as mp3 download via the Amchitka concert website. Using the money raised with the concert, the Don't Make a Wave Committee chartered a ship, the *Phyllis Cormack* owned and sailed by John Cormack. The ship was renamed *Greenpeace* for the protest after a term coined by activist Bill Darnell.

In the fall of 1971 the ship sailed towards Amchitka and faced the U.S. Coast Guard ship *Confidence* which forced the activists to turn back. Because of this and the increasingly bad weather the crew decided to return to Canada only to find out that the news about their journey and reported support from the crew of the *Confidence* had generated sympathy for their protest. After this Greenpeace tried to navigate to the test site with other vessels, until the U.S. detonated the bomb. The nuclear test was criticized and the U.S. decided not to continue with their test plans at Amchitka. In 1972, The Don't Make a Wave committee changed their official name to *Greenpeace Foundation*. While the organization was founded under a different name in 1970 and was officially named Greenpeace in 1972, the organization itself dates its birth to the first protest of 1971. Greenpeace also states that "there was no single founder, and the name, idea, spirit, tactics, and internationalism of the organization all can be said to have separate lineages".

As Rex Weyler put it in his chronology, *Greenpeace*, in 1969, Irving and Dorothy Stowe's "quiet home on Courtenay Street would soon become a hub of monumental, global significance". Some of the first Greenpeace meetings were held there, and it served as the first office of the Greenpeace Foundation. After the office in the Stowe home, (and after the first concert fund-raiser) Greenpeace functions moved to other private homes before settling, in the fall of 1974, in a small office shared with the SPEC environmental group, at 2007 W. 4th Avenue, at Maple Street, across from the Bimini neighbourhood pub. The address of this office has since been changed to 2009 W. 4th Avenue. The building still exists, and the office is up the stair at the "2009" door.

First Campaigns after Amchitka

After the nuclear tests at Amchitka were over, Greenpeace moved its focus to the French atmospheric nuclear weapons testing at the Moruroa Atoll in French Polynesia. The young organization needed help for their protests and were contacted by David McTaggart, a former businessman living in New Zealand. In 1972 the yacht *Vega*, a 12.5-metre (41 ft) ketch owned by David McTaggart, was renamed *Greenpeace III* and sailed in an anti-nuclear protest into the exclusion zone at Moruroa to attempt to disrupt French nuclear testing. This voyage was sponsored and organized by the New Zealand branch of the Campaign for Nuclear Disarmament. The French Navy tried to stop the protest in several ways, including assaulting David McTaggart. McTaggart was supposedly beaten to the point that he lost sight in one of his eyes. Luckily, one of McTaggart's crew members photographed the incident and went public. After the assault was publicized, France announced it would stop the atmospheric nuclear tests.

In the mid-1970s some Greenpeace members started an independent campaign, Project Ahab, against commercial whaling, since Irving Stowe was against Greenpeace focusing on other issues than nuclear weapons. After Irving Stowe died in 1975, Phyllis Cormack left from Vancouver to face Soviet whalers on the coast of California. Greenpeace activists disrupted the whaling by going between the harpoons and the whales, and the footage of the protests spread across the world. Later in the 1970s the organization widened its focus to include toxic waste and commercial seal hunting.

Organizational Development

Greenpeace evolved from a group of Canadian protesters in a sail boat, into a less conservative group of environmentalists who were more reflective of the counterculture and hippie youth movements of the 1960s and 1970s. The social and cultural background from which Greenpeace emerged heralded a period of de-conditioning away from old world antecedents and sought to develop new codes of social, environmental and political behaviour. Historian Frank Zelko has commented that "unlike Friends of the Earth, for example, which sprung fully formed from the forehead of David Brower, Greenpeace developed in a more evolutionary manner."

In the mid-1970s independent groups using the name Greenpeace started springing up world wide. By 1977 there were 15 to 20 Greenpeace groups around the world. At the same time the Canadian Greenpeace office was heavily in debt. Disputes between offices over fund-raising and organizational direction split the global movement as the North American offices were reluctant to be under the authority of the Vancouver office and its president Patrick Moore.

After the incidents of Moruroa, David McTaggart had moved to France to battle in court with the French state and helped to develop the cooperation of European Greenpeace groups. David McTaggart lobbied the Canadian Greenpeace Foundation to accept a new structure which would bring the scattered Greenpeace offices under the auspices of a single global organization.

The European Greenpeace paid the debt of the Canadian Greenpeace office and on October 14, 1979, Greenpeace International came into existence. Under the new structure, the local offices would contribute a percentage of their income to the international organization, which would take responsibility for setting the overall direction of the movement with each regional office having one vote. Some Greenpeace groups, namely London Greenpeace (dissolved in 2001) and the US-based Greenpeace Foundation (still operational) however decided to remain independent from Greenpeace International.

Priorities and Campaigns

On its official website, Greenpeace defines its mission as the following:

Greenpeace is an independent global campaigning organization that acts to change attitudes and behaviour, to protect and conserve the environment and to promote peace by:

- Catalysing an energy revolution to address the number one threat facing our planet: climate change.

- Defending our oceans by challenging wasteful and destructive fishing, and creating a global network of marine reserves.
- Protecting the world's remaining ancient forests which are depended on by many animals, plants and people.
- Working for disarmament and peace by reducing dependence on finite resources and calling for the elimination of all nuclear weapons.
- Creating a toxin free future with safer alternatives to hazardous chemicals in today's products and manufacturing.
- Campaigning for sustainable agriculture by encouraging socially and ecologically responsible farming practices. —*Greenpeace International*

Climate and Energy

Greenpeace was one of the first parties to formulate a sustainable development scenario for climate change mitigation, which it did in 1993. According to sociologists Marc Mormont and Christine Dasnoy, Greenpeace played a significant role in raising public awareness of global warming in the 1990s. The organization has also focused on CFCs, because of both their global warming potential and their effect on the ozone layer. Greenpeace was one of the leading participants advocating early phase-out of ozone depleting substances in the Montreal Protocol. In the early 1990s, Greenpeace developed a CFC-free refrigerator technology, "Greenfreeze" for mass production together with the refrigerator industry. United Nations Environment Programme awarded Greenpeace for "outstanding contributions to the protection of the Earth's ozone layer" in 1997. In 2007 one third of the world's total production of refrigerators were based on Greenfreeze technology, with over 200 million units in use.

Currently Greenpeace considers global warming to be the greatest environmental problem facing the Earth. Greenpeace calls for global greenhouse gas emissions to peak in 2015 and to decrease as close to zero as possible by 2050. For this Greenpeace calls for the industrialized countries to cut their emissions at least 40% by 2020 (from 1990 levels) and to give substantial funding for developing countries to build a sustainable energy capacity, to adapt to the inevitable consequences of global warming, and to stop deforestation by 2020. Together with EREC, Greenpeace has formulated a global energy scenario, "Energy Revolution", where 80% of the world's total energy is produced with renewables, and the emissions of the energy sector are decreased by over 80% of the 1990 levels by 2050.

Using direct action, Greenpeace has protested several times against coal by occupying coal power plants and blocking coal shipments and mining operations, in places such as New Zealand, Svalbard, Australia, and the United Kingdom. Greenpeace is also critical of extracting petroleum from oil sands and has used direct action to block the oil sand operations at Athabasca, Canada.

The Kingsnorth Court Case

In October 2007, six Greenpeace protesters were arrested for breaking in to the Kingsnorth power station, climbing the 200 metre smokestack, painting the name Gordon on the chimney, and causing an estimated £30,000 damage. At their subsequent trial they admitted trying to shut the station down, but argued that they were legally justified because they were trying to prevent climate change from causing greater damage to property elsewhere around the world. Evidence was heard from David Cameron's environment adviser Zac Goldsmith, climate scientist James E. Hansen and an Inuit leader from Greenland, all saying that climate change was already seriously affecting life around the world. The six activists were acquitted. It was the first case where preventing property damage caused by climate change has been used as part of a "lawful excuse" defence in court. Both The Daily Telegraph and The Guardian described the acquittal as embarrassment to the Brown Ministry. In December 2008 *The New York Times* listed the acquittal in its annual list of the most influential ideas of the year.

"Go Beyond Oil"

As part of their stance on the subject re-newable energy, Greenpeace have launched the "Go Beyond Oil" campaign. The campaign is focused on slowing, and eventually ending, the world's consumption of oil; with activist activities taking place against companies that pursue oil drilling as a venture. Much of the activities of the "Go Beyond Oil" campaign have been focused on drilling for oil in the Arctic and areas affected by the Deepwater Horizon disaster. The activities of Greenpeace in the arctic have mainly involved the Edinburgh-based oil and gas exploration company, Cairn Energy; and range from protests at the Cairn Energy's headquarters to scalling their oil rigs in an attempt to halt the drilling process.

The "Go Beyond Oil" campaign also involves applying political pressure on the governments who allow oil exploration in their territories; with the group stating that one of the key aims of the "Go Beyond Oil" campaign is to "work to expose the lengths the oil industry is willing to go to squeeze the last barrels out of the ground and put pressure on industry and governments to move beyond oil."

Nuclear Power

Greenpeace views nuclear power as a relatively minor industry with major problems, such as environmental damage and risks from uranium mining, nuclear weapons proliferation, and unresolved questions concerning nuclear waste.

The organization argues that the potential of nuclear power to mitigate global warming is marginal, referring to the IEA energy scenario where an increase in world's nuclear capacity from 2608 TWh in 2007 to 9857 TWh by 2050 would cut global greenhouse gas emissions less than 5% and at require 32 nuclear reactor units of 1000MW capacity built per year until 2050.

According to Greenpeace the slow construction times, construction delays, and hidden costs, all limit the mitigation potential of nuclear power. This makes the IEA scenario technically and financially unrealistic. They also argue that binding massive amounts of investments on nuclear energy would take funding away from more effective solutions. Greenpeace views the construction of Olkiluoto 3 nuclear power plant in Finland as an example of the problems on building new nuclear power.

Anti-nuclear Advertisement

In 1994, Greenpeace published an anti-nuclear newspaper advert which included a claim that nuclear facilities Sellafield would kill 2000 people in the next 10 years, and an image of a hydrocephalus-affected child said to be a victim of nuclear weapons testing in Kazakhstan. Advertising Standards Authority viewed the claim concerning Sellafield unsubstantiated, and ASA did not accept that the child's condition was caused by radiation. This resulted in banning of the advert. Greenpeace did not admit fault, stating that a Kazakhstan doctor had said that the child's condition was due to nuclear testing. Adam Woolf from Greenpeace also stated that, "fifty years ago there were many experts who would be lined up and swear there was no link between smoking and bad health." The UN has estimated that the nuclear weapon tests in Kazakhstan caused about 100,000 people to suffer over three generations.

Press Release Blunder

In Philadelphia, in 2006, Greenpeace issued a press release that said "In the twenty years since the Chernobyl tragedy, the world's worst nuclear accident, there have been nearly [FILL IN ALARMIST AND ARMAGEDDONIST FACTOID HERE]," The final report warned of plane crashes and reactor meltdowns. According to a Greenpeace spokesman, the memo was a joke that was accidentally released.

Forest Campaign

Greenpeace aims at protecting intact primary forests from deforestation and degradation with the target of zero deforestation by 2020. Greenpeace has accused several corporations, such as Unilever, Nike, and McDonald's of having links to the deforestation of the tropical rainforests, resulting in policy changes in several of the companies under criticism. Greenpeace, together with other environmental NGOs, also campaigned for ten years for the EU to ban import of illegal timber. The EU decided to ban illegal timber on July 2010. As deforestation contributes to global warming, Greenpeace has demanded that REDD (Reduced Emission from Deforestation and Forest Degradation) should be included in the climate treaty following the Kyoto treaty.

Removal of Ancient Tree

In June 1995, Greenpeace took a trunk of a tree from the forests of the proposed national park of Koitajoki in Ilomantsi, Finland and put it on display at exhibitions held in Austria and Germany. Greenpeace said in a press conference that the tree was originally from a logged area in the ancient forest which was supposed to be protected. Metsähallitus accused Greenpeace of theft and said that the tree was from a normal forest and had been left standing because of its old age. Metsähallitus also said that the tree had actually crashed over a road during a storm. The incident received publicity in Finland, for example in the large newspapers *Helsingin Sanomat* and *Ilta-Sanomat*. Greenpeace replied that the tree had fallen down because of the protective forest around it had been clearcut, and that they wanted to highlight the fate of old forests in general, not the fate of one particular tree. Greenpeace also highlighted that Metsähallitus admitted the value of the forest afterwards as Metsähallitus currently refers to Koitajoki as a distinctive area because of its old growth forests.

The 'Tokyo Two'

In 2008, two Greenpeace anti-whaling activists, Junichi Sato and Toru Suzuki, stole a case of whale meat from a delivery depot in Aomori prefecture, Japan. Their intention was to expose what they considered embezzlement of the meat collected during whale hunts. After a brief investigation of their allegations was ended, Sato and Suzuki were arrested and charged with theft and trespass. Amnesty International said that the arrests and following raids on Greenpeace Japan office and homes of five of Greenpeace staff members were aimed at intimidating activists and non-governmental organizations. They were convicted of theft and trespass in September 2010 by the Aomori district court.

Genetically Modified Organisms (GMOs)

Greenpeace has also supported the rejection of GM food from the US in famine-stricken Zambia as long as supplies of non-genetically engineered grain exist, stating that the US "should follow in the European Union's footsteps and allow aid recipients to choose their food aid, buying it locally if they wish. This practise can stimulate developing economies and creates more robust food security", adding that, "if Africans truly have no other alternative, the controversial GE maize should be milled so it can't be planted. It was this condition that allowed Zambia's neighbours Zimbabwe and Malawi to accept it." After Zambia banned all GM food aid, the former agricultural minister of Zambia criticized, "how the various international NGOs that have spoken approvingly of the government's action will square the body count with their various consciences." Concerning the decision of Zambia, Greenpeace has stated that, "it was obvious to us that if no non-GM aid was being offered then they should absolutely accept GM food aid. But the Zambian government decided to refuse the GM food. We offered our opinion to the Zambian government and, as many governments do, they disregarded our advice."

Greenpeace on Golden Rice

Greenpeace opposes the planned use of golden rice, a variety of *Oryza sativa* rice produced through genetic engineering to biosynthesize beta-carotene, a precursor of pro-vitamin A in the edible parts of rice. According to Greenpeace, golden rice has not managed to do anything about malnutrition for 10 years during which alternative methods are already tackling malnutrition.

The alternative proposed by Greenpeace is to discourage mono-cropping and to increase production of crops which are naturally nutrient-rich (containing other nutrients not found in golden rice in addition to beta-carotene). Greenpeace argues that resources should be spent on programs that are already working and helping to relieve malnutrition. The Golden Rice Project acknowledges that, "While the most desirable option is a varied and sufficient diet, this goal is not always achievable, at least not in the short term."

The renewal of these concerns coincided with the publication of a paper in the journal *Nature* about a version of golden rice with much higher levels of beta carotene. This "golden rice 2" was developed and patented by Syngenta, which provoked Greenpeace to renew its allegation that the project is driven by profit motives. Dr. C.S. Prakash, who is the director of the Center for Plant Biotechnology Research at Tuskegee University and is president of the AgBioWorld Foundation expressed the opinion that, "critics condemned biotechnology as something that is purely for profit, that is being pursued only in the West, and with no benefits to the consumer. Golden Rice proves them wrong, so they need to discredit it any way they can."

Although Greenpeace had admitted efficiency to be its primary concern, as early as 2001, statements from March and April 2005 also continued to express concern over human health and environmental safety. Greenpeace has opposed releasing golden rice to fields as opposed to farming in greenhouses, which according to golden rice developer Ingo Potrykus, limits the amount of material needed for human safety testing.

Toxics

In July 2011, Greenpeace released its Dirty Laundry report accusing some of the world's top sportswear brands of releasing toxic waste into China's rivers. The report profiles the problem of water pollution resulting from the release of toxic chemicals associated with the country's textile industry. Investigations focused on wastewater discharges from two facilities in China; one belonging to the Youngor Group located on the Yangtze River Delta and the other to Well

Dyeing Factory Ltd. located on a tributary of the Pearl River Delta. Scientific analysis of samples from both facilities revealed the presence of hazardous and persistent hormone disruptor chemicals, including alkylphenols, perfluorinated compounds and perfluorooctane sulfonate.

The report goes on to assert that the Youngor Group and Well Dyeing Factory Ltd. - the two companies behind the facilities - have commercial relationships with a range of major clothing brands, including Abercrombie & Fitch, Adidas, Bauer Hockey, Calvin Klein, Converse (shoe company), Cortefiel, H&M, Lacoste, Li Ning (company), Metersbonwe Group, Nike, Phillips-Van Heusen and Puma AG.

Organizational Structure

Governance: Greenpeace consists of *Greenpeace International* (officially Stichting Greenpeace Council) based in Amsterdam, Netherlands, and 28 regional offices operating in 45 countries. The regional offices work largely autonomously under the supervision of Greenpeace International. The executive director of Greenpeace is elected by the board members of Greenpeace International. The current director of Greenpeace International is Kumi Naidoo and the current Chair of the Board is Lalita Ramdas. Greenpeace has a staff of 2,400 and 15,000 volunteers globally.

Each regional office is led by a regional executive director elected by the regional board of directors. The regional boards also appoint a trustee to The Greenpeace International Annual General Meeting, where the trustees elect or remove the board of directors of Greenpeace International. The role of the annual general meeting is also to discuss and decide the overall principles and strategically important issues for Greenpeace in collaboration with the trustees of regional offices and Greenpeace International board of directors.

Funding

Greenpeace receives its funding from individual supporters and foundations. Greenpeace screens all major donations in order to ensure it does not receive unwanted donations. The organization does not accept money from governments, intergovernmental organizations, political parties or corporations in order to avoid their influence. Donations from foundations which are funded by political parties or receive most of their funding from governments or intergovernmental organizations are rejected. Foundation donations are also rejected if the foundations attach unreasonable conditions, restrictions or constraints on Greenpeace activities or if the donation would compromise the independence and aims of Greenpeace. Since in the mid-1990s the number of supporters started to decrease, Greenpeace pioneered the use of face-to-face fundraising where fundraisers actively seek new supporters at public places, subscribing them for a monthly direct debit donation. In 2008, most of the €202.5 million received by the organization was donated by about 2.6 million regular supporters, mainly from Europe.

In September 2003, the Public Interest Watch (PIW) complained to the Internal Revenue Service, claiming that Greenpeace USA tax returns were inaccurate and in violation of the law. PIW charged that Greenpeace was using non-profit donations for advocacy instead of charity and educational purposes. PIW asked the IRS to investigate the complaint. Greenpeace rejected the accusations and challenged PIW to disclose its funders, a request rejected by then-Executive Director of PIW, Mike Hardiman, because PIW does not have 501c3 tax exempt status like Greenpeace does in the U.S. The IRS conducted an extensive review and concluded in December 2005 that Greenpeace USA continued to qualify for its tax-exempt status. In March 2006 *The Wall Street Journal* reported that PIW had been funded by ExxonMobil prior to PIW's request to investigate Greenpeace.

Ships

Since Greenpeace was founded, seagoing ships have played a vital role in its campaigns. Once the *Rainbow Warrior III* is completed (expected in 2011), the group will have three ocean-going ships, the *Esperanza, Arctic Sunrise* and *Rainbow Warrior III.*

The First Rainbow Warrior

In 1978, Greenpeace launched the original *Rainbow Warrior*, a 40-metre (130 ft), former fishing trawler named for the Cree legend that inspired early activist Robert Hunter on the first voyage to Amchitka. Greenpeace purchased the *Rainbow Warrior* (originally launched as the *Sir William Hardy* in 1955) at a cost of £40,000. Volunteers restored and refitted it over a period of four months. First deployed to disrupt the hunt of the Icelandic whaling fleet, the *Rainbow Warrior* would quickly become a mainstay of Greenpeace campaigns. Between 1978 and 1985, crew members also engaged in direct action against the ocean-dumping of toxic and radioactive waste, the Grey Seal hunt in Orkney and nuclear testing in the Pacific. Japan's Fisheries

Agency has labelled Greenpeace ships as "anti-whaling vessels" and "environmental terrorists". In May 1985, the vessel was instrumental for 'Operation Exodus', the evacuation of about 300 Rongelap Atoll islanders whose home had been contaminated with nuclear fallout from a US nuclear test two decades ago which had never been cleaned up and was still having severe health effects on the locals.

Later in 1985 the *Rainbow Warrior* was to lead a flotilla of protest vessels into the waters surrounding Moruroa atoll, site of French nuclear testing. The sinking of the Rainbow Warrior occurred when the French government secretly bombed the ship in Auckland harbour on orders from François Mitterrand himself.

This killed Dutch freelance photographer Fernando Pereira, who thought it was safe to enter the boat to get his photographic material after a first small explosion, but drowned as a result of a second, larger explosion.

The attack was a public relations disaster for France after it was quickly exposed by the New Zealand police. The French Government in 1987 agreed to pay New Zealand compensation of NZ$13 million and formally apologised for the bombing. The French Government also paid £ 2.3 million compensation to the family of the photographer.

The Second Rainbow Warrior

In 1989 Greenpeace commissioned a replacement vessel, also named the Rainbow Warrior (also referred as *Rainbow Warrior II*), which was retired from service on the 16th of August 2011 to be replaced by the third Rainbow Warrior.

In 2005 the *Rainbow Warrior II* ran aground on and damaged the Tubbataha Reef in the Philippines while inspecting the reef for coral bleaching. Greenpeace was fined US$7,000 for damaging the reef and agreed to pay the fine saying they felt responsible for the damage, although Greenpeace stated that the Philippines government had given it outdated charts. The park manager of Tubbataha appreciated the quick action Greenpeace took to assess the damage to the reef.

.

Other Vessels

Along with the *Rainbow Warriors*, Greenpeace has had several other ships in its service: MV *Sirius*, MV *Solo*, MV *Greenpeace*, MV *Arctic Sunrise* and MV *Esperanza*, the last two being in service today.

Reactions and Responses to Greenpeace Activities

Lawsuits have been filed against Greenpeace for lost profits, reputation damage and "sailor mongering". In 2004 it was revealed that the Australian government was willing to offer a subsidy to Southern Pacific Petroleum on the condition that the oil company would take legal action against Greenpeace, which had campaigned against the Stuart Oil Shale Project.

Some corporations, such as Royal Dutch Shell, BP and Électricité de France have reacted to Greenpeace campaigns by spying on Greenpeace activities and infiltrating Greenpeace offices. Greenpeace activists have also been targets of phone tapping, death threats, violence and even state terrorism in the case of bombing of the *Rainbow Warrior*.

Local Greenpeaces

Greenpeace Chile: In Chile, the organization is affiliated as "Greenpeace Chile" was founded in 1981 and is a government recognized NGO there.

Greenpeace East Asia

Greenpeace East Asia's first China office was opened in Hong Kong in 1997. Headquartered in Beijing, the office now campaigns in Mainland China, Hong Kong, Taiwan and South Korea.

4

Oil Production Plant

An oil production plant (sometimes called an oil terminal) is a facility which performs processing of production fluids from oil wells in order to separate out key components and prepare them for export. This is distinct from an oil depot, which do not have processing facilities.

Typical production fluids are a mixture of oil, gas and produced water. Many permanent offshore platforms have full oil production facilities onboard. Smaller platforms and subsea wells must export raw production fluid to the nearest production facility, which may be on a nearby offshore processing platform or an onshore terminal.

The production plant is said to begin after the production wing valve on the oil well. The product from each well is piped through the choke valve, which regulates the rate of flow. The flowlines are gathered in a manifold and routed into a separator, which will gravity separate the three components. Once the oil has been separated from the gas and produced water, it is usually routed to a coalescer before being metered and pumped to the onshore terminal.

The produced water is often routed to a hydrocyclone to remove entrained oil and solids and then either re-injected into the reservoir or dumped overboard depending on the circumstances and cleanliness of the water. The associated gas is initially dubbed “wet gas” as it is saturated with water and liquid alkanes. The gas is typically routed through scrubbers, compressors and coolers which will remove the bulk of the liquids. This “dry gas” may be exported, re-injected into the reservoir, used for gas lift, flared or used as fuel for the installation’s power generators. Onshore terminals generally have fired heaters followed by separators and coalescers to stabilise the crude and remove any produced water not separated offshore. Onshore separators tend to operate at a lower pressure than the offshore separators and so more gas is evolved. The associated gas is generally compressed, dew-pointed and exported via a dedicated pipeline. If gas export is uneconomical then it may be flared. Onshore terminals frequently have large crude oil storage tanks to allow offshore production to continue if the export route becomes unavailable. Export to the refinery is either by pipeline or tanker.

Oil-storage Trade

The oil-storage trade is a trading strategy where oil tank owners and companies that lease storage buy oil for immediate delivery and hold it in their storage tanks, then sell contracts for future delivery at a higher price. When delivery dates approach, they close out existing contracts and sell new ones for future delivery of the same oil.

The oil never moves out of storage. Trading in this fashion is only successful if the forward market is in "contango", that is if the price of oil in the future also known as forward prices are higher than current prices or spot prices. Storing oil became big business in 2008 and 2009, with many participants - including Wall Street giants such as Morgan Stanley, Goldman Sachs or Citicorp - turning sizeable profits simply by sitting on tanks of oil.

It has been estimated that one in twelve of the largest oil tankers are being used for the storage, rather than transportation of oil, and that if lined up end to end, the tankers would stretch out for 26 miles.

Oil Refinery

An oil refinery or petroleum refinery is an industrial process plant where crude oil is processed and refined into more useful petroleum products, such as gasoline, diesel fuel, asphalt base, heating oil, kerosene, and liquefied petroleum gas. Oil refineries are typically large sprawling industrial complexes with extensive piping running throughout, carrying streams of fluids between large chemical processing units.

In many ways, oil refineries use much of the technology of, and can be thought of as types of chemical plants. The crude oil feedstock has typically been processed by an oil production plant. There is usually an oil depot (tank farm) at or near an oil refinery for storage of bulk liquid products. An oil refinery is considered an essential part of the downstream side of the petroleum industry.

***Figure:** Anacortes Refinery (Tesoro), on the north end of March Point southeast of Anacortes, Washington*

Operation

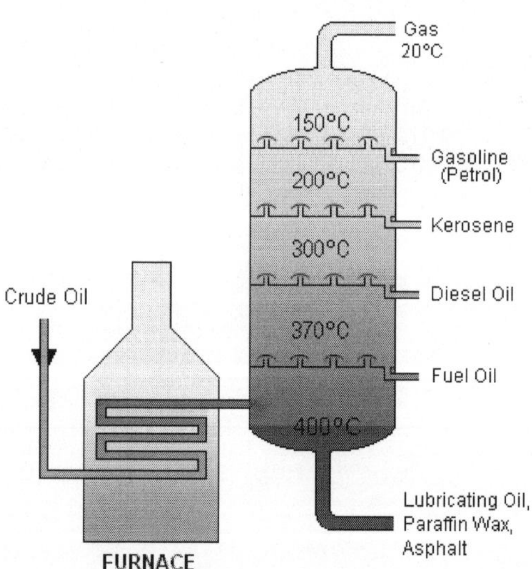

***Figure:** Crude oil is separated into fractions by fractional distillation. The fractions at the top of the fractionating column have lower boiling points than the fractions at the bottom. The heavy bottom fractions are often cracked into lighter, more useful products. All of the fractions are processed further in other refining units.*

Raw or unprocessed crude oil is not generally useful. Although "light, sweet" (low viscosity, low sulfur) crude oil has been used directly as a burner fuel for steam vessel propulsion, the lighter elements form explosive vapours in the fuel tanks and are therefore hazardous, especially in warships. Instead, the hundreds of different hydrocarbon molecules in crude oil are separated in a refinery into components which can be used as fuels, lubricants, and as feedstock in petrochemical processes that manufacture such products as plastics, detergents, solvents, elastomers and fibers such as nylon and polyesters.

Petroleum fossil fuels are burned in internal combustion engines to provide power for ships, automobiles, aircraft engines, lawn mowers, chainsaws, and other machines. Different boiling points allow the hydrocarbons to be separated by distillation. Since the lighter liquid products are in great demand for use in internal combustion engines, a modern refinery will convert heavy hydrocarbons and lighter gaseous elements into these higher value products.

Oil can be used in a variety of ways because it contains hydrocarbons of varying molecular masses, forms and lengths such as paraffins, aromatics, naphthenes (or cycloalkanes), alkenes, dienes, and alkynes. While the molecules in crude oil include different atoms such as sulfur and nitrogen, the hydrocarbons are the most common form of molecules, which are molecules of varying lengths and complexity made of hydrogen and carbon atoms, and a small number of oxygen atoms. The differences in the structure of these molecules account for their varying physical and chemical properties, and it is this variety that makes crude oil useful in a broad range of applications.

Once separated and purified of any contaminants and impurities, the fuel or lubricant can be sold without further processing. Smaller molecules such as isobutane and propylene or butylenes can be recombined to meet specific octane requirements by processes such as alkylation, or less commonly, dimerization.

Octane grade of gasoline can also be improved by catalytic reforming, which involves removing hydrogen from hydrocarbons producing compounds with higher octane ratings such as aromatics. Intermediate products such as gasoils can even be reprocessed to break a heavy, long-chained oil into a lighter short-chained one, by various forms of cracking such as fluid catalytic cracking, thermal cracking, and hydrocracking. The final step in gasoline production is the blending of fuels with different octane ratings, vapour pressures, and other properties to meet product specifications.

Oil refineries are large scale plants, processing about a hundred thousand to several hundred thousand barrels of crude oil a day. Because of the high capacity, many of the units operate continuously, as opposed to processing in batches, at steady state or nearly steady state for months to years.

The high capacity also makes process optimization and advanced process control very desirable.

Major Products

Petroleum products are usually grouped into three categories: light distillates (LPG, gasoline, naphtha), middle distillates (kerosene, diesel), heavy distillates and residuum (heavy fuel oil, lubricating oils, wax, asphalt). This classification is based on the way crude oil is distilled and separated into fractions (called distillates and residuum) as in the above drawing.

- Liquified petroleum gas (LPG)
- Gasoline (also known as petrol)
- Naphtha
- Kerosene and related jet aircraft fuels
- Diesel fuel
- Fuel oils
- Lubricating oils
- Paraffin wax
- Asphalt and tar
- Petroleum coke.

Oil refineries also produce various intermediate products such as hydrogen, light hydrocarbons, reformate and pyrolysis gasoline. These are not usually transported but instead are blended or processed further on-site. Chemical plants are thus often adjacent to oil refineries. For example, light hydrocarbons are steam-cracked in an ethylene plant, and the produced ethylene is polymerized to produce polyethene.

Common Process units Found in a Refinery

- Desalter unit washes out salt from the crude oil before it enters the atmospheric distillation unit.

- Atmospheric distillation unit distills crude oil into fractions.
- Vacuum distillation unit further distills residual bottoms after atmospheric distillation.
- Naphtha hydrotreater unit uses hydrogen to desulfurize naphtha from atmospheric distillation. Must hydrotreat the naphtha before sending to a Catalytic Reformer unit.
- Catalytic reformer unit is used to convert the naphtha-boiling range molecules into higher octane reformate (reformer product). The reformate has higher content of aromatics and cyclic hydrocarbons). An important byproduct of a reformer is hydrogen released during the catalyst reaction. The hydrogen is used either in the hydrotreaters or the hydrocracker.
- Distillate hydrotreater unit desulfurizes distillates (such as diesel) after atmospheric distillation.
- Fluid catalytic cracker (FCC) unit upgrades heavier fractions into lighter, more valuable products.
- Hydrocracker unit uses hydrogen to upgrade heavier fractions into lighter, more valuable products.
- Visbreaking unit upgrades heavy residual oils by thermally cracking them into lighter, more valuable reduced viscosity products.
- Merox unit treats LPG, kerosene or jet fuel by oxidizing mercaptans to organic disulfides.
- Coking units (delayed coking, fluid coker, and flexicoker) process very heavy residual oils into gasoline and diesel fuel, leaving petroleum coke as a residual product.
- Alkylation unit produces high-octane component for gasoline blending.
- Dimerization unit converts olefins into higher-octane gasoline blending components. For example, butenes can be dimerized into isooctene which may subsequently be hydrogenated to form isooctane. There are also other uses for dimerization.
- Isomerization unit converts linear molecules to higher-octane branched molecules for blending into gasoline or feed to alkylation units.
- Steam reforming unit produces hydrogen for the hydrotreaters or hydrocracker.

- Liquified gas storage units store propane and similar gaseous fuels at pressure sufficient to maintain them in liquid form. These are usually spherical vessels or bullets (horizontal vessels with rounded ends.
- Storage tanks store crude oil and finished products, usually cylindrical, with some sort of vapor emission control and surrounded by an earthen berm to contain spills.
- Slug catcher used when product (crude oil and gas) that comes from a pipeline with two-phase flow, has to be buffered at the entry of the units.
- Amine gas treater, Claus unit, and tail gas treatment convert hydrogen sulfide from hydrodesulfurization into elemental sulfur.
- Utility units such as cooling towers circulate cooling water, boiler plants generates steam, and instrument air systems include pneumatically operated control valves and an electrical substation.
- Wastewater collection and treating systems consist of API separators, dissolved air flotation (DAF) units and further treatment units such as an activated sludge biotreater to make water suitable for reuse or for disposal.
- Solvent refining units use solvent such as cresol or furfural to remove unwanted, mainly aromatics from lubricating oil stock or diesel stock.
- Solvent dewaxing units remove the heavy waxy constituents petrolatum from vacuum distillation products.

Flow Diagram of Typical Refinery

The diagram depicts only one of the literally hundreds of different oil refinery configurations. The diagram also does not include any of the usual refinery facilities providing utilities such as steam, cooling water, and electric power as well as storage tanks for crude oil feedstock and for intermediate products and end products.

There are many process configurations other than that depicted above. For example, the vacuum distillation unit may also produce fractions that can be refined into endproducts such as: spindle oil used in the textile industry, light machinery oil, motor oil, and steam cylinder oil. As another example, the vacuum residue may be processed in a coker unit to produce petroleum coke.

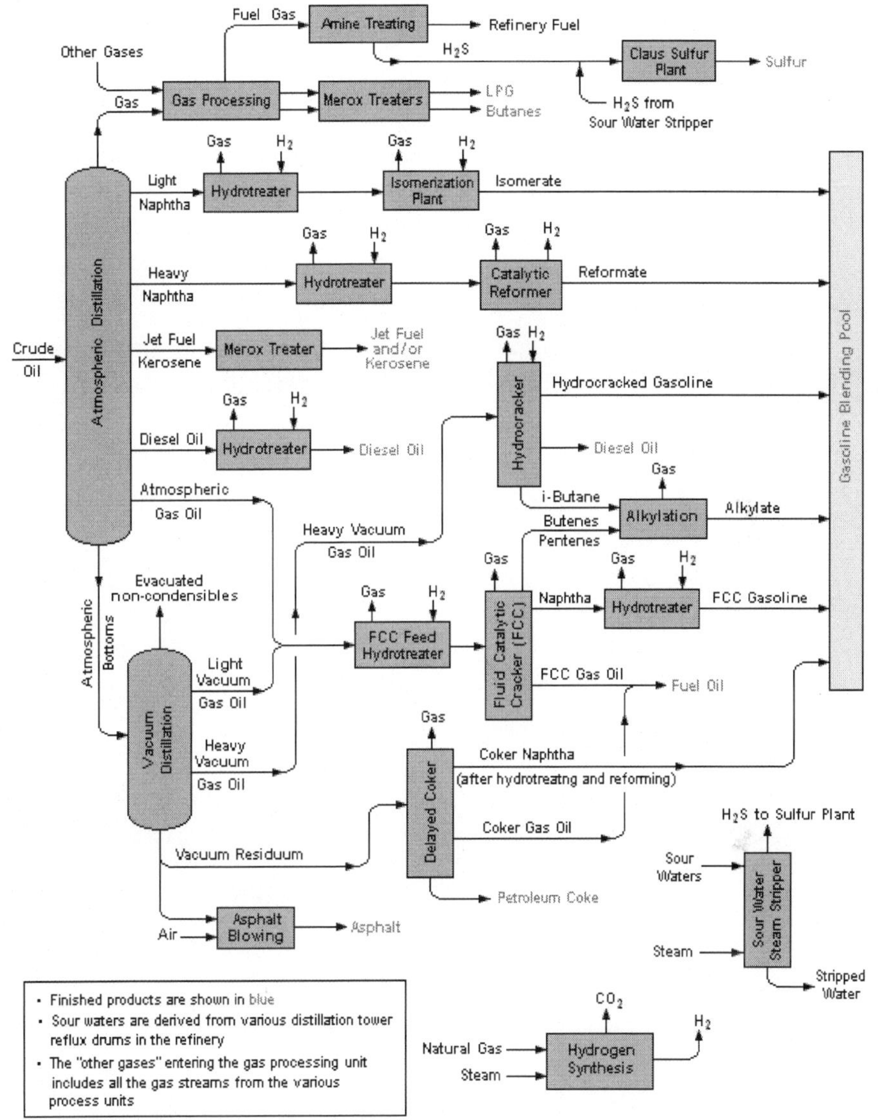

Figure: *Schematic flow diagram of a typical oil refinery*

Specialty End Products

These will blend various feedstocks, mix appropriate additives, provide short term storage, and prepare for bulk loading to trucks, barges, product ships, and railcars.

- Gaseous fuels such as propane, stored and shipped in liquid form under pressure in specialized railcars to distributors.

- Liquid fuels blending (producing automotive and aviation grades of gasoline, kerosene, various aviation turbine fuels, and diesel fuels, adding dyes, detergents, antiknock additives, oxygenates, and anti-fungal compounds as required). Shipped by barge, rail, and tanker ship. May be shipped regionally in dedicated pipelines to point consumers, particularly aviation jet fuel to major airports, or piped to distributors in multi-product pipelines using product separators called pipeline inspection gauges ("pigs").
- Lubricants (produces light machine oils, motor oils, and greases, adding viscosity stabilizers as required), usually shipped in bulk to an offsite packaging plant.
- Wax (paraffin), used in the packaging of frozen foods, among others. May be shipped in bulk to a site to prepare as packaged blocks.
- Sulfur (or sulfuric acid), byproducts of sulfur removal from petroleum which may have up to a couple percent sulfur as organic sulfur-containing compounds. Sulfur and sulfuric acid are useful industrial materials. Sulfuric acid is usually prepared and shipped as the acid precursor oleum.
- Bulk tar shipping for offsite unit packaging for use in tar-and-gravel roofing.
- Asphalt unit. Prepares bulk asphalt for shipment.
- Petroleum coke, used in specialty carbon products or as solid fuel.
- Petrochemicals or petrochemical feedstocks, which are often sent to petrochemical plants for further processing in a variety of ways. The petrochemicals may be olefins or their precursors, or various types of aromatic petrochemicals.

Siting/Locating of Petroleum Refineries

A party searching for a site to construct a refinery or a chemical plant needs to consider the following issues:

- The site has to be reasonably far from residential areas.
- Infrastructure should be available for supply of raw materials and shipment of products to markets.
- Energy to operate the plant should be available.
- Facilities should be available for waste disposal.

Refineries which use a large amount of steam and cooling water need to have an abundant source of water. Oil refineries therefore are often located nearby navigable rivers or on a sea shore, nearby a port. Such location also gives access to transportation by river or by sea. The advantages of transporting crude oil by pipeline are evident, and oil companies often transport a large volume of fuel to distribution terminals by pipeline. Pipeline may not be practical for products with small output, and rail cars, road tankers, and barges are used.

Petrochemical plants and solvent manufacturing (fine fractionating) plants need spaces for further processing of a large volume of refinery products for further processing, or to mix chemical additives with a product at source rather than at blending terminals.

Safety and Environmental Concerns

Figure: *Fire at Union Oil refinery, Wilmington, California, 1951*

Figure: *MiRO refinery at Karlsruhe*

The refining process releases numerous different chemicals into the atmosphere; consequently, there are substantial air pollution emissions and a notable odor normally accompanies the presence of a refinery. Aside from air pollution impacts there are also wastewater concerns, risks of industrial accidents such as fire and explosion, and noise health effects due to industrial noise. The public has demanded that many governments place restrictions on contaminants that refineries release, and most refineries have installed the equipment needed to comply with the requirements of the pertinent environmental protection regulatory agencies. In the United States, there is strong pressure to prevent the development of new refineries, and no major refinery has been built in the country since Marathon's Garyville, Louisiana facility in 1976.

However, many existing refineries have been expanded during that time. Environmental restrictions and pressure to prevent construction of new refineries may have also contributed to rising fuel prices in the United States. Additionally, many refineries (over 100 since the 1980s) have closed due to obsolescence and/or merger activity within the industry itself. This activity has been reported to Congress and in specialized studies not widely publicised.

Environmental and safety concerns mean that oil refineries are sometimes located some distance away from major urban areas. Nevertheless, there are many instances where refinery operations are close to populated areas and pose health risks such as in the Campo de Gibraltar, a CEPSA refinery near the towns of Gibraltar, Algeciras, La Linea, San Roque and Los Barrios with a combined population of over 300,000 residents within a 5-mile (8.0 km) radius and the CEPSA refinery in Santa Cruz on the island of Tenerife, Spain which is sited in a densely populated city centre and next to the only two major evacuation routes in and out of the city. In California's Contra Costa County and Solano County, a shoreline necklace of refineries, built in the early 20th century before this area was populated, and associated chemical plants are adjacent to urban areas in Richmond, Martinez, Pacheco, Concord, Pittsburg, Vallejo and Benicia, with occasional accidental events that require "shelter in place" orders to the adjacent populations.

Corrosion Problems and Prevention

Petroleum refineries run as efficiently as possible to reduce costs. One major factor that decreases efficiency is corrosion of the metal components found throughout the process line of the hydrocarbon refining process. Corrosion causes the failure of parts in addition to dictating the cleaning schedule of the refinery, during which the entire production facility must be shut down and cleaned. The cost of corrosion in the petroleum industry has been estimated at US$3.7 billion.

Corrosion occurs in various forms in the refining process, such as pitting corrosion from water droplets, embrittlement from hydrogen, and stress corrosion cracking from sulfide attack. From a materials standpoint, carbon steel is used for upwards of 80% of refinery components, which is beneficial due to its low cost. Carbon steel is resistant to the most common forms of corrosion, particularly from hydrocarbon impurities at temperatures below 205 °C, but other corrosive chemicals and environments prevent its use everywhere. Common replacement materials are low alloy steels containing chromium and molybdenum, with stainless steels containing more chromium dealing with more corrosive environments. More expensive materials commonly used are nickel, titanium, and copper alloys. These are primarily saved for the most problematic areas where extremely high temperatures or very corrosive chemicals are present.

Corrosion is fought by a complex system of monitoring, preventative repairs and careful use of materials. Monitoring methods include both off-line checks taken during maintenance and on-line monitoring. Off-line checks measure corrosion after it has occurred, telling the engineer when equipment must be replaced based on the historical information he has collected. This is referred to as preventative management.

On-line systems are a more modern development, and are revolutionizing the way corrosion is approached. There are several types of on-line corrosion monitoring technologies such as linear polarization resistance, electrochemical noise and electrical resistance. On-Line monitoring has generally had slow reporting rates in the past (minutes or hours) and been limited by process conditions and sources of error but newer technologies can report rates up to twice per minute with much higher accuracy (referred to as real-time monitoring).

This allows process engineers to treat corrosion as another process variable that can be optimized in the system. Immediate responses to process changes allow the control of corrosion mechanisms, so they can be minimized while also maximizing production output. In an ideal situation having on-line corrosion information that is accurate and real-time will allow conditions that cause high corrosion rates to be identified and reduced. This is known as predictive management.

Materials methods include selecting the proper material for the application. In areas of minimal corrosion, cheap materials are preferable, but when bad corrosion can occur, more expensive but longer lasting materials should be used. Other materials methods come in the form of protective barriers between corrosive substances and the equipment metals. These can be either a lining of refractory material such as standard Portland cement or other special acid-resistant cements that are shot onto the inner surface of the vessel. Also available are thin overlays of more expensive metals that protect cheaper metal against corrosion without requiring lots of material.

History

The first oil refineries in the world were built by Ignacy £ukasiewicz near Jas³o, Poland from 1854 to 1856, but they were initially small as there was no real demand for refined fuel. As £ukasiewicz's kerosene lamp gained popularity, the refining industry grew in the area.

The world's first large refinery opened at Cîmpina, Romania, in 1856-1857, with United States investment. After being taken over by Nazi Germany, the Cîmpina refineries were bombed in Operation Tidal Wave by the Allies during the Oil Campaign of World War II. Another early large refinery is Oljeon, Sweden (1875) (Swedish name means *The Petroleum Isle*), now preserved as a museum near Engelsberg Ironworks, a UNESCO World Heritage Site, and part of the Ekomuseum Bergslagen.

At one point, the refinery in Ras Tanura, Saudi Arabia owned by Saudi Aramco was claimed to be the largest oil refinery in the world. For most of the 20th century, the largest refinery was the Abadan Refinery in Iran. This refinery suffered extensive damage during the Iran-Iraq war. The world's largest refinery complex is the Jamnagar Refinery Complex, consisting of two refineries side by side operated by Reliance Industries Limited in Jamnagar, India with a combined production capacity of 1,240,000 barrels per day (197,000 m^3/d) (J-1 660,000 bbl/d (105,000 m^3/d), J-2 580,000 bbl/d (92,000 m^3/d). PDVSA's Paraguana refinery complex in Venezuela with a capacity of 956,000 bbl/d (152,000 m^3/d) and SK Energy's Ulsan in South Korea with 840,000 bbl/d (134,000 m^3/d) are the second and third largest, respectively.

Oil Refining in the United States

In the 19th century, refineries in the U.S. processed crude oil primarily to recover the kerosene. There was no market for the more volatile fraction, including gasoline, which was considered waste and was often dumped directly into the nearest river. The invention of the automobile shifted the demand to gasoline and diesel, which remain the primary refined products today. Today, national and state legislation requires refineries to meet stringent air and water cleanliness standards. In fact, oil companies in the U.S. perceive obtaining a permit to build a modern refinery to be so difficult and costly that no new refineries have been built (though many have been expanded) in the U.S. since 1976.

More than half the refineries that existed in 1981 are now closed due to low utilization rates and accelerating mergers. As a result of these closures total US refinery capacity fell between 1981 to 1995, though the operating capacity stayed fairly constant in that time period at around 15,000,000 barrels per day (2,400,000 m^3/d). Increases in facility size and improvements in efficiencies have offset much of the lost physical capacity of the industry. In 1982 (the earliest data provided), the United States operate 301 refineries with a combined capacity of 17.9 million barrels (2,850,000 m^3) of crude oil each calendar day. In 2010, there were 149 operable U.S. refineries with a combined capacity of 17.6 million barrels (2,800,000 m^3) per calendar day.

In 2009 through 2010, as revenue streams in the oil business dried up and profitability of oil refineries fell due to lower demand for product and high reserves of supply preceding the economic recession, oil companies began to close or sell refineries. Due to EPA regulations, the costs associated with closing a refinery are very high, meaning that many former refineries are re-purposed.

K Factor Crude Oil Refining

In the oil and gas engineering industry, the K-factor formula is used to calculate pressure drop across fittings or a set of fittings in a piping system. K-factor, when multiplied by ρv2/2, gives pressure drop across the fittings. K-factor can also be used to find out the equivalent length of all the fittings in a piping system. When the total pressure drop across a line (straight line + fittings) is divided by pressure drop per unit length for that line, the result is equivalent straight length of that piping system accounting for all the bends and fittings.

List of Oil Refineries

This is a list of oil refineries. The Oil and Gas Journal also publishes a worldwide list of refineries annually in a country-by-country tabulation that includes for each refinery: location, crude oil daily processing capacity, and the size of each process unit in the refinery. For the U.S., the refinery list is further categorized state-by-state. The list usually appears in one of their December issues. It is about 45 pages in length and is updated each year with additions, deletions, name changes, capacity changes, etc.

World's largest refineries

Name of Refinery	*Location*	*Barrels per Day*
Jamnagar Refinery (Reliance Industries Limited)	Jamnagar, Gujarat, India	1,300,000
Paraguana Refinery Complex (PDVSA)	Paraguana, Falcon, Venezuela	940,000
Baytown Refinery (ExxonMobil)	Baytown, TX, USA	572,500
Baton Rouge Refinery (ExxonMobil)	Baton Rouge, LA, USA	503,000
Hovensa LLC (PDVSA) (Hess Corporation)	Virgin Islands, USA	500,000
Marathon Petroleum Company	Garyville LA, USA	464,000
BP Texas City	Texas City,TX USA	460,000
Abadan Refinery	Iran	450,000
BP Whiting Refinery	Whiting, IN	440,000
Citgo Petroleum Corporation	Lake Charles LA, USA	427,800
Royal Dutch Shell Pernis Refinery	Netherlands	416,000
Fawley Southampton Refinery (ExxonMobil)	Southampton, United Kingdom	347,000
Kirishi Refinery (Surgutneftegas)	Kirishi, Russia	337,000
Flint Hills Resources	Corpus Christi TX, USA	288,000
Chevron Richmond Refinery	Richmond CA, USA	240,000
Saudi Aramco Yanbu Refinery	Yanbu, KSA	235,000

Africa

Algeria:

- Adrar Refinery (CNPC), 12,500 bbl/d (1,990 m^3/d)
- Alger Refinery (Sonatrach), 55,000 bbl/d (8,700 m^3/d)
- Arzew Refinery (Sonatrach), 50,000 bbl/d (7,900 m^3/d)
- Hassi Messaoud Refinery (Sonatrach), 25,000 bbl/d (4,000 m^3/d)
- Skikda Refinery (Sonatrach), 356,000 bbl/d (56,600 m^3/d).

Angola:

- Cabinda Refinery (Chevron Corporation), 16,000 bbl/d (2,500 m^3/d)
- Luanda Refinery (Sonangol), 56,000 bbl/d (8,900 m^3/d)

Cameroon:

- Limbé Refinery (SoNaRa), 42,600 bbl/d (6,770 m^3/d).

Chad:

- N'Djamena Refinery (Société de Raffinage de N'Djamena), 20,000 bbl/d (3,200 m^3/d).

Congo:

- Pointe Noire Refinery (CORAF), 21,000 bbl/d (3,300 m^3/d).

Côte d'Ivoire:

- Abidjan Refinery (SIR), 68,000 bbl/d (10,800 m^3/d)
- Abidjan Bitumen Refinery (SMB), 10,000 bbl/d (1,600 m^3/d).

Egypt:

- Alexandria Ameriya Refinery (EGPC), 81,000 bbl/d (12,900 m^3/d)
- Alexandria El Mex Refinery (EGPC), 117,000 bbl/d (18,600 m^3/d)
- Alexandria MIDOR Refinery (EGPC), 100,000 bbl/d (16,000 m^3/d)
- Asyut Refinery (EGPC), 47,000 bbl/d (7,500 m^3/d)
- Cairo Mostorod Refinery (EGPC), 142,000 bbl/d (22,600 m^3/d)
- El Nasr Refinery (EGPC), 132,000 bbl/d (21,000 m^3/d)
- El Suez Refinery (EGPC), 70,000 bbl/d (11,000 m^3/d)
- Tanta Refinery (EGPC), 35,000 bbl/d (5,600 m^3/d)
- Wadi Feran Refinery (EGPC), 8,550 bbl/d (1,359 m^3/d).

Gabon:

- Port Gentil Refinery (Sogara), 25,000 bbl/d (4,000 m^3/d).

Ghana:

- Tema Oil Refinery (TOR), 45,000 bbl/d (7,200 m^3/d).

Kenya:

- Mombasa Refinery (KPRL), 70,000 bbl/d (11,000 m^3/d).

Libya:

- Zawiya Refinery (NOC), 120,000 bbl/d (19,000 m^3/d)
- Ra's Lanuf Refinery (NOC), 220,000 bbl/d (35,000 m^3/d)
- El-Brega Refinery (NOC), 10,000 bbl/d (1,600 m^3/d)
- Sarir Refinery (AGOC), 10,000 bbl/d (1,600 m^3/d)
- Tobruk Refinery (AGOC), 20,000 bbl/d (3,200 m^3/d).

Morocco:

- Mohammedia Refinery (SAMIR), 127,000 bbl/d (20,200 m^3/d)

Nigeria:

- Kaduna Refinery (NNPC), 110,000 bbl/d (17,000 m^3/d)
- Port Harcourt Refinery (NNPC), 210,000 bbl/d (33,000 m^3/d)
- Warri Refinery (NNPC), 125,000 bbl/d (19,900 m^3/d).

Senegal:

- Dakar Refinery (SAR), 127,000 bbl/d (20,200 m^3/d).

South Africa:

- Cape Town Refinery (Chevron Corporation), 110,000 bbl/d (17,000 m^3/d)
- Durban Engen Refinery (Petronas), 122,000 bbl/d (19,400 m^3/d)
- Durban Sapref Refinery (Sapref), 125,000 bbl/d (19,900 m^3/d)
- Sasolburg Refinery (Sasol), 125,000 bbl/d (19,900 m^3/d).

Sudan:

- El Obeid Refinery (ERC), 15,000 bbl/d (2,400 m^3/d)
- Khartoum Refinery (KRC), 100,000 bbl/d (16,000 m^3/d).

Tunisia:

- Bizerte Refinery (STIR), 34,000 bbl/d (5,400 m^3/d).

Zambia:

- Ndola Refinery (Indeni), 34,000 bbl/d (5,400 m^3/d).

Asia

Bangladesh:

- Eastern Refinery.

China:

- Fushun Petrochemical Refinery (CNPC), 160,000 bbl/d (25,000 m^3/d).

India:

- Assam
  - o Digboi Refinery, Assam (IOC), 13,000 bbl/d (2,100 m^3/d)
  - o Guwahati Refinery, Assam (IOC), 20,000 bbl/d (3,200 m^3/d)
  - o Bongaigaon Refinery, (IOC), 48,000 bbl/d (7,600 m^3/d)
  - o Numaligarh Refinery Limited, Assam (NRL), 58,000 bbl/d (9,200 m^3/d)
- Bihar
  - o Barauni Refinery (IOC), 116,000 bbl/d (18,400 m^3/d)
- Punjab
  - o Guru Gobind Singh Refinery Bhatinda (HPCL and Mittal energy limited -HMEL)Template:14 MMTPA
- Gujarat
  - o Jamnagar Refinery (Reliance Industries), 1,240,000 bbl/d (197,000 m^3/d)
  - o Essar Refinery (Essar Oil), 300,000 bbl/d (48,000 m^3/d)
  - o Gujarat Refinery (IOC), 170,000 bbl/d (27,000 m^3/d)
- Haryana
  - o Panipat Refinery (IOC), 300,000 bbl/d (48,000 m^3/d)
- Karnataka
  - o Mangalore Refinery (MRPL), 199,000 bbl/d (31,600 m^3/d)
- Madhya Pradesh
  - o Bina Refinery (BORL), 116,000 bbl/d (18,400 m^3/d)
- West Bengal
  - o Haldia Refinery (IOC), 116,000 bbl/d (18,400 m^3/d)

- Uttar Pradesh
  - Mathura Refinery (IOC), 156,000 bbl/d (24,800 m^3/d)
- Maharashtra
  - Mumbai Refinery (HPCL), 107,000 bbl/d (17,000 m^3/d)
  - Mumbai Refinery Mahaul (BPCL), 135,000 bbl/d (21,500 m^3/d)
- Andhra Pradesh
  - Visakhapatnam Refinery (HPCL), 150,000 bbl/d (24,000 m^3/d)
  - Tatipaka Refinery (ONGC), 1,600 bbl/d (250 m^3/d)
- Kerala
  - Kochi Refinery (BPCL), 172,000 bbl/d (27,300 m^3/d)
- Tamil Nadu
  - Chennai Refinery (IOC), 185,000 bbl/d (29,400 m^3/d)
  - Nagapattnam Refinery (CPCL), 20,000 bbl/d (3,200 m^3/d).

Indonesia:

- Musi Refinery (Pertamina), 135,200 bbl/d (21,500 m^3/d)
- Balongan Refinery (Pertamina), 125,000 bbl/d (19,900 m^3/d)
- Dumai Refinery (Pertamina), 120,000 bbl/d (19,000 m^3/d)
- Cilacap Refinery (Pertamina), 348,000 bbl/d (55,300 m^3/d)
- Balikpapan Refinery (Pertamina), 260,000 bbl/d (41,000 m^3/d)
- Sungai Pakning Refinery (Pertamina), 50,000 bbl/d (7,900 m^3/d)
- Pangkalan Brandan Refinery (Pertamina), 5,000 bbl/d (790 m^3/d)
- Cepu Refinery (Pertamina), 3,800 bbl/d (600 m^3/d)
- Kasim Refinery (Pertamina), 10,000 bbl/d (1,600 m^3/d)
- Tuban Refinery (PT TPPI), 100,000 bbl/d (16,000 m^3/d).

Iran:

- Abadan Refinery (NIOC), 450,000 bbl/d (72,000 m^3/d)
- Arak Refinery (NIOC), 150,000 bbl/d (24,000 m^3/d)
- Tehran Refinery (NIOC), 225,000 bbl/d (35,800 m^3/d)
- Isfahan Refinery (NIOC), 265,000 bbl/d (42,100 m^3/d)
- Tabriz Refinery (NIOC), 112,000 bbl/d (17,800 m^3/d)
- Shiraz Refinery (NIOC), 40,000 bbl/d (6,400 m^3/d)

- Lavan Refinery (NIOC), 20,000 bbl/d (3,200 m^3/d)
- Bandar Abbas Refinery (NIOC), 232,000 bbl/d (36,900 m^3/d)
- Kermanshah refinery (NIOC),21,000 bpd.

Iraq:

- Basrah Refinery (INOC), 126,000 bbl/d (20,000 m^3/d)
- Daurah Refinery (INOC), 100,000 bbl/d (16,000 m^3/d)
- Kirkuk Refinery (INOC), 270,000 bbl/d (43,000 m^3/d)
- Baiji Salahedden Refinery (INOC), 140,000 bbl/d (22,000 m^3/d)
- Baiji North Refinery (INOC), 150,000 bbl/d (24,000 m^3/d)
- Khanaqin/Alwand Refinery (INOC), 10,500 bbl/d (1,670 m^3/d)
- Samawah Refinery (INOC), 27,000 bbl/d (4,300 m^3/d)
- Haditha Refinery (INOC), 14,000 bbl/d (2,200 m^3/d)
- Muftiah Refinery (INOC), 4,500 bbl/d (720 m^3/d)
- Gaiyarah Refinery (INOC), 4,000 bbl/d (640 m^3/d)
- Irbel Refinery (INOC), 40,000 bbl/d (6,400 m^3/d).

Japan:

- Chiba Refinery (Cosmo Oil) (Cosmo Oil), 240,000 bbl/d (38,000 m^3/d)
- Yokkaichi Refinery (Cosmo Oil), 175,000 bbl/d (27,800 m^3/d)
- Sakai Refinery (Cosmo Oil) (Cosmo Oil), 80,000 bbl/d (13,000 m^3/d)
- Sakaide Refinery (Cosmo Oil), 140,000 bbl/d (22,000 m^3/d)
- Muroran Refinery (Nippon Oil Corporation (NOC)), 180,000 bbl/d (29,000 m^3/d)
- Sendai Refinery (Nippon Oil Corporation (NOC)), 145,000 bbl/d (23,100 m^3/d)
- Negishi Yokahama Refinery (Nippon Oil Corporation (NOC)), 340,000 bbl/d (54,000 m^3/d)
- Osaka Refinery (Nippon Oil Corporation (NOC)) 115,000 bpd
- Mizushima Refinery (Nippon Oil Corporation (NOC)), 250,000 bbl/d (40,000 m^3/d)
- Marifu Refinery (Nippon Oil Corporation (NOC)) 127,000 bpd
- Toyama Refinery (Nihonkai Oil/Nippon Oil Corporation (NOC)), 60,000 bbl/d (9,500 m^3/d)

- Kubiki Refinery (Teikoku Oil), 4,410 bbl/d (701 m^3/d)
- Chiba Refinery (Kyokuto) (Kyokuto Petroleum/ExxonMobil), 175,000 bbl/d (27,800 m^3/d)
- Kawasaki Refinery (TonenGeneral Sekiyu/ExxonMobil), 335,000 bbl/d (53,300 m^3/d)
- Wakayama Refinery (TonenGeneral Sekiyu/ExxonMobil), 170,000 bbl/d (27,000 m^3/d)
- Sakai Refinery (TonenGeneral) (TonenGeneral Sekiyu/ ExxonMobil), 156,000 bbl/d (24,800 m^3/d)
- Nishihara Refinery (Nansei sekiyu/Petrobras), 100,000 bbl/d (16,000 m^3/d)
- Keihin Refinery (Toa Oil/Shell), 185,000 bbl/d (29,400 m^3/d)
- Showa Yokkaichi Refinery (Showa Yokkaichi/Shell), 210,000 bbl/d (33,000 m^3/d)
- Yamaguchi Refinery (Seibu Oil/Shell), 120,000 bbl/d (19,000 m^3/d)
- Sodegaura Refinery (Fuji Oil Campany), 192,000 bbl/d (30,500 m^3/d)
- Kashima Refinery (Kashima Oil Campany/Japan Energy), 210,000 bbl/d (33,000 m^3/d)
- Mizushima Refinery (Japan Energy) (Japan Energy), 205,200 bbl/d (32,620 m^3/d)
- Shikoku Refinery (Taiyo Oil), 120,000 bbl/d (19,000 m^3/d)
- Ohita Refinery (Kyusyu Oil), 160,000 bbl/d (25,000 m^3/d)
- Hokkaido Refinery (Idemitsu Kosan), 140,000 bbl/d (22,000 m^3/d)
- Chiba Refinery (Idemitsu) (Idemitsu Kosan), 220,000 bbl/d (35,000 m^3/d)
- Aichi Refinery (Idemitsu Kosan), 160,000 bbl/d (25,000 m^3/d)
- Tokuyama Refinery (Idemitsu Kosan), 120,000 bbl/d (19,000 m^3/d).

Jordan:

- Jordan Refinery, Zarqa, Az Zarqa, (Jordan Petroleum Refinery Company), 65,000 bbl/d (10,300 m^3/d).

Kazakhstan:

- Shymkent Refinery (PetroKazakhstan), 160,000 bbl/d (25,000 m^3/d)

- Pavlodar Refinery (KazMunayGas), 162,600 bbl/d (25,850 m^3/d)
- Atyrau Refinery (KazMunayGas), 104,400 bbl/d (16,600 m^3/d)

Kuwait:

- Mina Al-Ahmadi Refinery (KNPC), 470,000 bbl/d (75,000 m^3/d)
- Shuaiba Refinery (KNPC), 200,000 bbl/d (32,000 m^3/d)
- Mina Abdullah Refinery (KNPC), 270,000 bbl/d (43,000 m^3/d)

Malaysia:

- Melaka I Refinery (Petronas), 100,000 bbl/d (16,000 m^3/d)[3]
- Melaka II Refinery (Petronas/ConocoPhillips), 100,000 bbl/d (16,000 m^3/d)[4]
- Kertih Refinery (Petronas), 40,000 bbl/d (6,400 m^3/d)
- Port Dickson Refinery (Royal Dutch Shell), 156,000 bbl/d (24,800 m^3/d)[5]
- Lutong Refinery (Royal Dutch Shell), 45,000 bbl/d (7,200 m^3/d). Has been closed.
- Esso Port Dickson Refinery (ExxonMobil), 88,000 bbl/d (14,000 m^3/d).

Oman:

- Mina Al Fahal Oman Refinery Company (ORC) 85,000 bbl/d (13,500 m^3/d)
- Sohar Refinery Company (SRC) 116,000 bbl/d (18,400 m^3/d)
- Dukum Refinery Company (DRC) 200,000 bpd (Proposed).

Pakistan:

- six oil refrenies
- Pakistan Refinery Limited (PRL), 15,000 bbl/d (2,400 m^3/d)
- National Refinery Limited (NRL),64,000 bpd
- Attock Refinery Limited (ARL), 46,000 bbl/d (7,300 m^3/d)
- Enar Petroleum Refining Facility (EPRF), 3,000 bbl/d (480 m^3/d)
- TransAisa Refinery Limited (TRL), 96,000 bbl/d (15,300 m^3/d)
- Taba Refinery Limited (TRL), 76,000 bbl/d (12,100 m^3/d).

Papua New Guinea:

- InterOil Refinery, Port Moresby (InterOil), 32,500 bbl/d (5,170 m^3/d).

Philippines:

- Limay Refinery (Petron), 180,000 bbl/d (29,000 m^3/d)
- Tabangao Refinery (Royal Dutch Shell), 120,000 bbl/d (19,000 m^3/d)
- Batangas Refinery (Caltex(Chevron)), 86,000 bbl/d (13,700 m^3/d).

Qatar:

- Um Said Refinery (QP Refinery 100%), 147,000 bbl/d (23,400 m^3/d)
- Lafan Refinery (Qatar Petroleum 51%, ExxonMobil 10%, Total 10%, Idemitsu 10%, Cosmo 10%, Mitsui 4.5%, Marubeni 4.5%), 146,000 bbl/d (23,200 m^3/d)
- AL-Shahene Refinery (2012), 250,000 bbl/d (40,000 m^3/d).

Saudi Arabia:

- Riyadh Refinery (Saudi Aramco), 120,000 bbl/d (19,000 m^3/d)
- Rabigh Refinery (PetroRabigh), 400,000 bbl/d (64,000 m^3/d)
- Jeddah Refinery (Saudi Aramco), 100,000 bbl/d (16,000 m^3/d)
- Ras Tanura Refinery (Saudi Aramco), 550,000 bbl/d (87,000 m^3/d)
- Yanbu' Refinery (Saudi Aramco), 225,000 bbl/d (35,800 m^3/d)
- Yanbu' Refinery (SAMREF) (Saudi Aramco/Exxon Mobil), 400,000 bbl/d (64,000 m^3/d)
- Yanbu' Refinery (Saudi Aramco/ConocoPhillips), 400,000 bbl/d (64,000 m^3/d)
- Jubail Refinery (SASREF) (Saudi Aramco/Shell), 305,000 bbl/d (48,500 m^3/d)
- Jubail Refinery (Saudi Aramco/TOTAL), 400,000 bbl/d (64,000 m^3/d).

Singapore:

- ExxonMobil Jurong Island Refinery (ExxonMobil), 605,000 bbl/d (96,200 m^3/d)
- SRC Jurong Island Refinery (Singapore Refining Corporation), 285,000 bbl/d (45,300 m^3/d)
- Shell Pulau Bukom Refinery (Royal Dutch Shell), 458,000 bbl/d (72,800 m^3/d)

Sri Lanka:

- Sapugaskanda Oil Refinery [cpc], 51,000 bbl/d (8,100 m^3/d)

South Korea:

- SK Energy Co., Ltd. Ulsan Refinery (SK Energy), 850,000 bbl/d (135,000 m^3/d)
- S-Oil Ulsan Refinery (S-Oil), 560,000 bbl/d (89,000 m^3/d)
- GS-Caltex Yeosu Refinery (GS-Caltex), 730,000 bbl/d (116,000 m^3/d)
- SK Energy Co., Ltd. Inchon Refinery (SK Energy), 275,000 bbl/d (43,700 m^3/d)
- Hyundai Daesan Refinery (Hyundai), 275,000 bbl/d (43,700 m^3/d)

Taiwan:

- Talin Refinery (CPC), 100,000 bbl/d (16,000 m^3/d)
- Kaohsiung Refinery (CPC), 270,000 bbl/d (43,000 m^3/d)
- Taoyuan Refinery (CPC), 200,000 bbl/d (32,000 m^3/d)
- Mailiao Refinery (*Formosa*), 450,000 bbl/d (72,000 m^3/d).

Thailand:

- Thai Oil Refinery (Thai Oil Company of PTT), 220,000 bbl/d (35,000 m^3/d)
- IRPC Refinery (IRPC PLC of PTT), 215,000 bbl/d (34,200 m^3/d)
- Rayong Refinery (Rayong Refinery PLC of PTT), 145,000 bbl/d (23,100 m^3/d)
- SPRC Refinery (Star Petroleum Refining Company of PTT), 150,000 bbl/d (24,000 m^3/d)
- Bangchak Refinery (Bangchak Petroleum of PTT), 120,000 bbl/d (19,000 m^3/d)
- Sri Racha Refinery (ExxonMobil), 170,000 bbl/d (27,000 m^3/d)
- Rayong Purifier Refinery (Rayong Purifier Company), 17,000 bbl/d (2,700 m^3/d)

Turkmenistan:

- Seidi, 120,000 bbl/d (19,000 m^3/d)
- Turkmenbashi, 116,000 bbl/d (18,400 m^3/d)

United Arab Emirates:

- Al-Ruwais Refinery (Abu Dhabi Oil Refining Company), 280,000 bbl/d (45,000 m^3/d)
- Umm Al-Narr Refinery (Abu Dhabi Oil Refining Company), 90,000 bbl/d (14,000 m^3/d)
- Jebel Ali Refinery (ENOC), 120,000 bbl/d (19,000 m^3/d)
- Hamriyah Sharjah Refinery (Sharjah Oil), 71,300 bbl/d (11,340 m^3/d)

Vietnam:

- Dung Quat Refinery (Petrovietnam), 148,000 bbl/d (23,500 m^3/d)

Yemen:

- Aden Refinery, (Aden Refinery Company), 120,000 bbl/d (19,000 m^3/d)
- Marib Refinery, (Yemen Hunt Oil Company), 10,000 bbl/d (1,600 m^3/d).

Europe

Austria:

- Schwechat Refinery, (OMV), 176,000 bbl/d (28,000 m^3/d)

Azerbaijan:

- Haydar Aliev Refinery (SOCAR), 220,000 bbl/d (35,000 m^3/d)
- Azerineftyag Refinery (SOCAR), 239,000 bbl/d (38,000 m^3/d)

Bosnia and Herzegovina:

- Bosanski Brod Refinery.

Belarus:

- Mozyr Refinery, (Slavneft),[7] 95,000 bbl/d (15,100 m^3/d)
- Novopolotsk Refinery, (Naftan), [8] 88,000 bbl/d (14,000 m^3/d)

Belgium:

- Total Antwerp Refinery, (Total), 360,000 bbl/d (57,000 m^3/d)
- ExxonMobil Antwerp Refinery, (ExxonMobil), 333,000 bbl/d (52,900 m^3/d)
- Antwerp N.V. Refinery, (Vitol), 35,000 bbl/d (5,600 m^3/d)
- BRC Antwerp (Petroplus), 115,000 bbl/d (18,300 m^3/d).

Bulgaria:

- LUKOIL Neftochim Burgas, (LUKOIL), 208,000 bbl/d (33,100 m^3/d)

Croatia:

- Rijeka Refinery, (INA), 90,000 bbl/d (14,000 m^3/d)
- Sisak Refinery, (INA), 60,000 bbl/d (9,500 m^3/d)

Czech Republic:

- Kralupy Refinery, (Ceska Rafinerska), 55,000 bbl/d (8,700 m^3/d)
- Litvinov Refinery, (Ceska Rafinerska), 120,000 bbl/d (19,000 m^3/d)
- Pardubice Refinery, (PARAMO), 15,000 bbl/d (2,400 m^3/d).

Denmark:

- Kalundborg Refinery, (Statoil), 110,000 bbl/d (17,000 m^3/d)
- Fredericia Refinery, (Royal Dutch Shell), 68,000 bbl/d (10,800 m^3/d).

Finland:

- Porvoo Refinery, (Neste Oil Oyj), 206,000 bbl/d (32,800 m^3/d)
- Naantali Refinery, (Neste Oil Oyj), 58,000 bbl/d (9,200 m^3/d)

France:

- Provence Refinery, (Total), 155,000 bbl/d (24,600 m^3/d)
- Normandy Refinery, (Total), 350,000 bbl/d (56,000 m^3/d)
- Flandres Refinery, (Total), 160,000 bbl/d (25,000 m^3/d) going to be closed
- Donges Refinery, (Total), 231,000 bbl/d (36,700 m^3/d).
- Feyzin Refinery, (Total), 119,000 bbl/d (18,900 m^3/d)
- Grandpuits Refinery, (Total), 99,000 bbl/d (15,700 m^3/d)
- Port Jérôme-Gravenchon Refinery, (ExxonMobil), 270,000 bbl/d (43,000 m^3/d)
- Fos-sur-Mer Refinery, (ExxonMobil), 140,000 bbl/d (22,000 m^3/d)
- Reichstett Refinery, (Petroplus), 77,000 bbl/d (12,200 m^3/d)
- Petit Couronne Refinery, (Petroplus), 142,000 bbl/d (22,600 m^3/d)
- Berre L'Etang Refinery, (LyondellBasell), 80,000 bbl/d (13,000 m^3/d)
- Lavera Marseilles Refinery, (Ineos), 220,000 bbl/d (35,000 m^3/d)
- Fort de France Refinery, (Total), 17,000 bbl/d (2,700 m^3/d).

Germany:

- Schwedt Refinery (PCK Raffinerie(Shell/PDVSA/BP/AET), 210,000 bbl/d (33,000 m^3/d)

- Ingolstadt Refinery (Bayernoil(OMV/Agip/PDVSA/BP)), 262,000 bbl/d (41,700 m^3/d)
- Ingolstadt Refinery (Petroplus), 110,000 bbl/d (17,000 m^3/d)
- Ruhr Ol Refinery (Rosneft/BP), 266,000 bbl/d (42,300 m^3/d)
- Buna SOW Leuna Refinery (Total), 222,000 bbl/d (35,300 m^3/d)
- Wilhelmshaven Refinery (ConocoPhillips), 300,000 bbl/d (48,000 m^3/d)
- Rheinland Werk Godorf Cologne Refinery (Royal Dutch Shell), 190,000 bbl/d (30,000 m^3/d)
- Rheinland Werk Wesseling Cologne Refinery (Royal Dutch Shell), 160,000 bbl/d (25,000 m^3/d)
- Mineralolraffinerie Karlsruhe Refinery (MiRo(Shell/ExxonMobil/PDVSA/BP/Conoco)) 285,000 bpd
- Burghausen Refinery (OMV) 70,000 bpd
- Mitteldeutschland Spergau Refinery (Total) 227,000 bpd
- Emsland Lingen Refinery (BP) 80,000 bpd
- Elbe Mineralolwerke Hamburg-Harburg Refinery (Royal Dutch Shell)
- Holborn Europa Raffinerie GmbH Hamburg (Holborn) 100,000 bpd.

Greece:

- Aspropyrgos Refinery, (Hellenic Petroleum), 135,000 bbl/d (21,500 m^3/d)
- Elefsina Refinery, (Hellenic Petroleum), 100,000 bbl/d (16,000 m^3/d)
- Thessaloniki Refinery, (Hellenic Petroleum), 66,500 bbl/d (10,570 m^3/d)

Hungary:

- Szazhalombatta Refinery, (MOL), 161,000 bbl/d (25,600 m^3/d)

Ireland:

- Whitegate Refinery, (ConocoPhillips), 71,000 bbl/d (11,300 m^3/d).

Italy:

- Esso Trecate, Novara Refinery, (ExxonMobil 74.1%/ERG 25.9%), 200,000 bbl/d (32,000 m^3/d)

- Esso Augusta Refinery, (ExxonMobil), 190,000 bbl/d (30,000 m^3/d)
- Sarroch Refinery, (Saras SPA), 300,000 bbl/d (48,000 m^3/d)
- Rome Refinery, (Total 77.5%/ERG 22.5%), 90,000 bbl/d (14,000 m^3/d)
- Falconara Marittima Ancona Refinery, (APIOIL), 85,000 bbl/d (13,500 m^3/d)
- Mantova Refinery, (IESItaliana), 55,000 bbl/d (8,700 m^3/d)
- Impianti Sud Refinery, (ISAB ERG), 214,000 bbl/d (34,000 m^3/d)
- Impianti Nord Refinery, (ISAB ERG), 160,000 bbl/d (25,000 m^3/d)
- Milazzo Refinery, (Eni/KNPC) 80,000 bpd
- Sannazzaro de' Burgondi Refinery, (Eni) 160,000 bpd
- Gela Refinery, (Eni) 100,000 bpd
- Taranto Refinery, (Eni) 90,000 bpd
- Livorno Refinery, (Eni) 84,000 bpd
- Porto Marghera Venice Refinery, (Eni) 70,000 bpd
- Cremona Refiney, (Tamoil) 80,000 bpd
- Iplom [9] Busalla, Genoa.

Lithuania:

- Mazeikiu Refinery, (Mazeikiu Nafta - PKN Orlen), 263,000 bbl/d (41,800 m^3/d)

Macedonia:

- OKTA Skopje Refinery, (Hellenic Petroleum), 50,000 bbl/d (7,900 m^3/d)

Netherlands:

- Shell Pernis Refinery, (Royal Dutch Shell), 416,000 bbl/d (66,100 m^3/d)
- Botlek (ExxonMobil) Rotterdam, 195,000 bbl/d (31,000 m^3/d)
- Total Refinery Netherlands - Vlissingen, (Total/LUKoil), 158,000 bbl/d (25,100 m^3/d)
- Europoort, (BP), 400,000 bbl/d (64,000 m^3/d)
- Q8-KPE Refinery Europoort, (Q8-Kuwait Petroleum Company), 80,000 bbl/d (13,000 m^3/d).

Norway:

- Slagen Refinery, (ExxonMobil), 110,000 bbl/d (17,000 m^3/d)
- Mongstad Refinery, (Statoil), 200,000 bbl/d (32,000 m^3/d).

Poland:

- Plock Refinery, (PKN Orlen), 276,000 bbl/d (43,900 m^3/d)
- Gdansk Refinery, (Grupa LOTOS), 210,000 bbl/d (33,000 m^3/d), (processing capacity after second distillation startup in 1Q2010).
- Czechowice Refinery, (Grupa LOTOS), 12,000 bbl/d (1,900 m^3/d), crude oil processing terminated 1Q2006.
- Trzebinia Refinery, (PKN Orlen), 4,000 bbl/d (640 m^3/d)
- Jaslo Oil Refinery , (Grupa LOTOS), 3,000 bbl/d (480 m^3/d), crude oil processing terminated 4Q2008.
- Jedlicze Refinery, (PKN Orlen), 2,800 bbl/d (450 m^3/d).
- Glimar Refinery, (Hudson Oil), 3,400 bbl/d (540 m^3/d), all operations (incl. crude oil processing) terminated 2005. Acquired 2011.

Portugal:

- Porto Refinery, (Galp Energia), 100,000 bbl/d (16,000 m^3/d)
- Sines Refinery, (Galp Energia), 200,000 bbl/d (32,000 m^3/d)

Romania:

- Arpechim Refinery Pite°ti, (Petrom/OMV), 70,000 bbl/d (11,000 m^3/d)
- Astra Refinery, (Interagro), closed for preservation , 20,000 bbl/d (3,200 m^3/d)
- Petrobrazi Refinery Ploie°ti, (Petrom/OMV), 90,000 bbl/d (14,000 m^3/d)
- Petromidia Constanþa Refinery, (Rompetrol), 100,000 bbl/d (16,000 m^3/d)
- Petrotel Lukoil Refinery Ploie°ti, (LUKOIL), 68,000 bbl/d (10,800 m^3/d)
- Petrolsub Suplacu de Barcãu Refinery, (Petrom/OMV), 15,000 bbl/d (2,400 m^3/d)
- RAFO One°ti, (Calder A), 70,000 bbl/d (11,000 m^3/d)

- Steaua Romanã Câmpina Refinery, (Omnimpex Chemicals), 15,000 bbl/d (2,400 m^3/d)
- Vega Ploie°ti Refinery, (Rompetrol), 20,000 bbl/d (3,200 m^3/d).

Kosovo:

- Kulla Exim Refinery, (KER), 101,000 bbl/d (16,100 m^3/d).

Russia:

Refineries with capacity more than 20,000 bbl/d (3,200 m^3/d)

Europe:

- Syzran Refinery, (Rosneft), 213,400 bbl/d (33,930 m^3/d)
- Novokuibyshevsk Refinery, (Rosneft), 191,500 bbl/d (30,450 m^3/d)
- Kuibyshev Oil Refinery, (Rosneft), 139,800 bbl/d (22,230 m^3/d)
- Salavatnefteorgsintez Refinery, (Gazprom, Salavat), 250,000 bbl/d (40,000 m^3/d)
- Volgograd Refinery, (LUKOIL), 193,000 bbl/d (30,700 m^3/d)
- Ukhta Refinery, (LUKOIL), 72,000 bbl/d (11,400 m^3/d)
- Perm Refinery, (LUKOIL), 235,000 bbl/d (37,400 m^3/d)
- NORSI-oil, (LUKOIL, Kstovo), 292,000 bbl/d (46,400 m^3/d)
- Ryazan Refinery, (TNK-BP), 253,000 bbl/d (40,200 m^3/d)
- Orsk Refinery, (Russneft), 159,000 bbl/d (25,300 m^3/d)
- Saratov Refinery, (TNK-BP), 108,000 bbl/d (17,200 m^3/d)
- Moscow Refinery, (Gazprom Neft/Central Fuel Company/ Tatneft), 213,000 bbl/d (33,900 m^3/d)
- Kirishi Refinery, (Surgutneftegas), 337,000 bbl/d (53,600 m^3/d)
- YaNOS Yaroslavl Refinery, (Slavneft), 132,000 bbl/d (21,000 m^3/d)
- Krasnodar Refinery, (Russneft), 58,000 bbl/d (9,200 m^3/d)
- Tuapse Refinery, (Rosneft), 85,000 bbl/d (13,500 m^3/d)
- Nizhnekamsk Refinery, (TAIF), 14,000 bbl/d (2,200 m^3/d)
- Ufa Refinery, (Bashneft), 190,000 bbl/d (30,000 m^3/d)
- Novo-Ufa Refinery, (Bashneft), 380,000 bbl/d (60,000 m^3/d)
- Ufaneftekhim Refinery, (Bashneft), 250,000 bbl/d (40,000 m^3/d)

Asia:

- Achinsk Refinery, (Rosneft), 131,000 bbl/d (20,800 m^3/d)
- Angarsk Petrochemical Refinery, (Rosneft), 384,000 bbl/d (61,100 m^3/d)
- Khabarovsk Refinery, (Alliance), 85,000 bbl/d (13,500 m^3/d)
- Komsomolsk Refinery, (Rosneft), 120,000 bbl/d (19,000 m^3/d)
- Nizhnevartovsk Refinery, (TNK-BP), 25,100 bbl/d (3,990 m^3/d)
- Omsk Refinery, (Gazprom Neft), 380,000 bbl/d (60,000 m^3/d).

Serbia:

- Panèevo Refinery (Naftna Industrija Srbije),
- Novi Sad Refinery (Naftna Industrija Srbije),
- Hemco Refinery

Slovakia:

- Slovnaft Bratislava Refinery, (MOL), 110,000 bbl/d (17,000 m^3/d)
- Petrochema Dubova Refinery, (russian investors),

Spain:

- Bilbao Refinery, (Repsol YPF), 220,000 bbl/d (35,000 m^3/d)
- Puertollano Refinery, (Repsol YPF), 140,000 bbl/d (22,000 m^3/d)
- Tarragona Refinery, (Repsol YPF), 160,000 bbl/d (25,000 m^3/d)
- A Coruña Refinery, (Repsol YPF), 120,000 bbl/d (19,000 m^3/d)
- Cartagena Refinery, (Repsol YPF), 220,000 bbl/d (35,000 m^3/d)
- Tenerife Refinery, (CEPSA), 90,000 bbl/d (14,000 m^3/d)
- Palos de la Frontera Refinery, (CEPSA), 100,000 bbl/d (16,000 m^3/d)
- Gibraltar-San Roque Refinery, (CEPSA), 240,000 bbl/d (38,000 m^3/d)
- Castellon Refinery, (BP), 100,000 bbl/d (16,000 m^3/d).

Switzerland:

- Cressier Refinery, (Petroplus), 68,000 bbl/d (10,800 m^3/d)
- Collombey-Muraz Refinery, (Tamoil), 45,000 bbl/d (7,200 m^3/d).

Turkey:

- Central Anatolian Refinery, (Tüpra°), 100,000 bbl/d (16,000 m^3/d)
- Izmit Refinery, (Tüpra°), 226,000 bbl/d (35,900 m^3/d)
- Aliaga Refinery, (Tüpra°), 200,000 bbl/d (32,000 m^3/d)
- Batman Refinery, (Tüpra°) 22,000 bbl/d (3,500 m^3/d)
- Akdeniz Refinery, (Petrol Ofisi) (still under construction)
- Dopu Akdeniz Petrol Refinery, (still under construction)
- Ýzmir Petrol Refinery, (Turcas-Socar) (still under construction)

Ukraine:

- Odessa Refinery, (LUKOIL), 70,000 bbl/d (11,000 m^3/d)
- LINOS Refinery, (TNK-BP), 320,000 bbl/d (51,000 m^3/d)
- Kherson Refinery, (Alliance), 36,000 bbl/d (5,700 m^3/d)
- Kremenchug Refinery, (Ukrtatnafta) 368,500 bpd
- Drogobych Refinery, (Pryvat) 40,000 bpd
- Neftekhimik Prikarpatya Nadvirna Refinery, (Pryvat) 39,000 bpd.

United Kingdom:

- Lindsey Oil Refinery, (Total), 223,000 bbl/d (35,500 m^3/d)
- Milford Haven Refinery, (Murco), 140,000 bbl/d (22,000 m^3/d)
- Pembroke Refinery, (Chevron), 220,000 bbl/d (35,000 m^3/d)
- Stanlow Refinery, (Essar Energy), 246,000 bbl/d (39,100 m^3/d)
- Teesside Refinery, (Petroplus), 117,000 bbl/d (18,600 m^3/d)
- Fawley Southampton Refinery, (ExxonMobil), 347,000 bbl/d (55,200 m^3/d)
- Humber Refinery, (ConocoPhillips), 221,000 bbl/d (35,100 m^3/d)
- Coryton Refinery, (Petroplus), 208,000 bbl/d (33,100 m^3/d)
- Grangemouth Refinery, (Ineos), 205,000 bbl/d (32,600 m^3/d)

North America

Aruba:

- Aruba Refinery (Valero) 275,000 bpd.

Canada:

Newfoundland and Labrador:

- North Atlantic Refinery, located in Come by Chance, (North Atlantic Refining), 115,000 bbl/d (18,300 m^3/d).

Nova Scotia:

- Imperial Oil Refinery - Dartmouth, (Imperial Oil), 89,000 bbl/d (14,100 m^3/d).

New Brunswick:

- Saint John, (Irving Oil), 300,000 bbl/d (48,000 m^3/d).

Quebec:

- Montreal-East, (Shell Canada), 161,000 bbl/d (25,600 m^3/d). Montreal East Refinery (Shell Canada). On June 4, 2010, Shell Canada officially announced the commencement to downgrade the refinery into a terminal, following the unsuccessful attempt to find a buyer to take over the plant.
- Montreal, (Suncor Energy), 160,000 bbl/d (25,000 m^3/d). Formerly Petro-Canada (before Aug 2009) and historically a Petrofina refinery. Montreal Refinery
- Montreal, Gulf Canada Oil, 70,000 bbl/d (11,000 m^3/d) Closed in 1985 and restarted in 2003. Montreal East Refinery (Gulf Oil Canada)
- Lévis, (Ultramar(Valero)), 215,000 bbl/d (34,200 m^3/d).

Ontario:

- Nanticoke Refinery, Nanticoke - (Imperial Oil), 112,000 bbl/d (17,800 m^3/d)
- Sarnia, (Imperial Oil), 115,000 bbl/d (18,300 m^3/d)
- Sarnia, (Suncor Energy), 85,000 bbl/d (13,500 m^3/d)
- Corunna, (Shell Canada), 72,000 bbl/d (11,400 m^3/d)

Lubricant Refinery:

- Mississauga, (Suncor Energy), 15,600 bbl/d (2,480 m^3/d) - aka Clarkson Refinery - base oil production is 13,600 bpd of API Group II capacity and 2,000 bpd of API Group III capacity. Formerly Petro-Canada (before Aug 2009) and historically a Gulf refinery.

Saskatchewan:

- CCRL Refinery Complex, Regina (Consumers' Co-operative Refineries Limited (CCRL)), 100,000 bbl/d (16,000 m^3/d)

Upgraders (improve the quality of crude for sale at a higher price)

- Husky Lloydminster Refinery, Lloydminster, (Husky Energy), 25,000 bbl/d (4,000 m^3/d)

- Husky Lloydminster Upgrader Lloydminster, (Husky Energy), 75,000 bbl/d (11,900 m^3/d).

Alberta:

- Strathcona Refinery, Edmonton, (Imperial Oil), 187,000 bbl/d (29,700 m^3/d)
- Scotford Refinery, Scotford, (Shell Canada), 100,000 bbl/d (16,000 m^3/d)
- Edmonton, (Suncor Energy), 135,000 bbl/d (21,500 m^3/d). Formerly Petro-Canada (before Aug 2009).
- Scotford Upgrader, Scotford, (AOSP - Shell Canada 60%, Chevron Corporation 20%, Marathon Oil 20%), 250,000 bpd (located next to Shell Refinery) raw bitumen
- Horizon Oil Sands, Fort McMurray, (Canadian Natural Resources Limited), 110,000 bbl/d (17,000 m^3/d) raw bitumen
- Long Lake, Fort McMurray, (OPTI Canada Inc. 35% and Nexen Inc. 65%), 70,000 bbl/d (11,000 m^3/d) raw bitumen
- Syncrude, Fort McMurray, (Canadian Oil Sands Trust, Imperial Oil, Suncor, Nexen, Conoco Phillips, Mocal Energy and Murphy Oil), 350,000 bbl/d (56,000 m^3/d) raw bitumen
- Suncor, Fort McMurray, (Suncor), 350,000 bbl/d (56,000 m^3/d) raw bitumen.

British Columbia:

- Burnaby Refinery, Burnaby, (Chevron Corporation), 52,000 bbl/d (8,300 m^3/d)
- Prince George Refinery, Prince George, (Husky Energy), 12,000 bbl/d (1,900 m^3/d)

Cuba:

- Nico Lopez Refinery (Cupet) 122,000 bpd
- Hermanos Diaz Refinery (Cupet) 102,500 bpd
- Cienfuegos Refinery (Cupet) 76,000 bpd.

Curaçao:

- Isla Refinery (PDVSA) 320,000 bpd.

Costa Rica:

- Puerto Limon Refinery (Recope), 8,000 bpd (start-up 1967) 2006, 25,000 bbl/d (4,000 m^3/d)

Dominican Republic:

- Haina Refinery (REFIDOMSA) 33,000 bpd (start-up 1973).

El Salvador:

- Refineria Petrolera de Acajutla S.A. de C.V. (RASA) (ExxonMobil) 22,000 bpd (start-up 1962).

Jamaica:

- Kingston Refinery (PCJ & PDVSA) 50,000 bpd.

Mexico:

- Reynosa Refinery (Pemex) Reynosa, Tamaulipas
- Minatitlan Refinery (Pemex) Minatitlan 167,000 bpd
- Cadereyta Refinery (Pemex) Cadereyta Jiménez, Nuevo Leon 217,000 bpd
- Tula Refinery (Pemex) Tula, Hidalgo 290,000 bpd
- Salamanca Refinery (Pemex) Salamanca, Guanajuato 192,000 bpd
- Ciudad Madero Refinery (Pemex) Ciudad Madero 152,000 bpd
- Salina Cruz Refinery (Pemex) Salina Cruz 227,000 bpd

Nicaragua:

- Cuesta del Plomo-Managua (ExxonMobil) 20,900 bpd (start-up 1962).

Trinidad and Tobago:

- Pointe-à-Pierre Refinery (Petrotrin) 165,000 bpd.

United States

Alabama:

- Tuscaloosa Refinery (Hunt Refining Company), Tuscaloosa 52,000 bbl/d (8,300 m^3/d)
- Saraland Refinery (Shell Oil Company), Saraland 80,000 bbl/d (13,000 m^3/d)
- Mobile Refinery (Gulf Atlantic Refining & Marketing), Mobile 16,700 bbl/d (2,660 m^3/d)

Alaska:

- Kenai Refinery (Tesoro), Kenai 72,000 bbl/d (11,400 m^3/d)
- Valdez Refinery (Petro Star), Valdez 50,000 bbl/d (7,900 m^3/d)

- North Pole Refinery (Petro Star), North Pole 17,000 bbl/d (2,700 m^3/d)
- Kuparuk Refinery (ConocoPhillips), Kuparuk 14,400 bbl/d (2,290 m^3/d).
- North Pole Refinery (Flint Hills Resources), North Pole 210,000 bbl/d (33,000 m^3/d)
- Prudhoe Bay Refinery (BP), Prudhoe Bay 12,500 bbl/d (1,990 m^3/d).

Arkansas:

- El Dorado Refinery (Lion Oil), El Dorado 70,000 bbl/d (11,000 m^3/d)
- Smackover Refinery (Cross Oil), Smackover 6,800 bbl/d (1,080 m^3/d).

California:

- Bakersfield Refinery (Alon USA), Bakersfield, 66,000 bbl/d (10,500 m^3/d)
- Bakersfield Refinery (Kern Oil), Bakersfield, 25,000 bbl/d (4,000 m^3/d)
- Bakersfield Refinery (San Joaquin Refining Company), Bakersfield, 24,300 bbl/d (3,860 m^3/d)
- Benicia Refinery (Valero), Benicia, 144,000 bbl/d (22,900 m^3/d)
- Carson Refinery (BP), Carson, 265,000 bbl/d (42,100 m^3/d)
- El Segundo Refinery (Chevron), El Segundo, 265,500 bbl/d (42,210 m^3/d)
- Golden Eagle Refinery (Tesoro), near Martinez, 166,000 bbl/d (26,400 m^3/d)
- Long Beach Refinery (Alon USA), Long Beach, 26,000 bbl/d (4,100 m^3/d)
- Martinez Refinery (Shell Oil Company), Martinez, 154,900 bbl/d (24,630 m^3/d)
- Oxnard Refinery (Tenby Inc), Oxnard, 2,800 bbl/d (450 m^3/d)
- Paramount Refinery (Paramount Petroleum), Paramount, 50,000 bbl/d (7,900 m^3/d)
- Richmond Refinery (Chevron), Richmond, 242,901 bbl/d (38,618.2 m^3/d)
- Rodeo San Francisco Refinery (ConocoPhillips), Rodeo, 100,000 bbl/d (16,000 m^3/d)

- Santa Maria Refinery (ConocoPhillips), Santa Maria, 41,800 bbl/d (6,650 m^3/d).
- Santa Maria Refinery (Greka Energy), Santa Maria, 9,500 bbl/d (1,510 m^3/d)
- South Gate Refinery (Lunday Thagard), South Gate, 8,500 bbl/d (1,350 m^3/d)
- Torrance Refinery (ExxonMobil), Torrance, 149,000 bbl/d (23,700 m^3/d)
- Wilmington Asphalt Refinery (Valero), Wilmington, 5,900 bbl/d (940 m^3/d)
- Wilmington Refinery (Tesoro), Wilmington, 133,100 bbl/d (21,160 m^3/d)
- Wilmington Refinery (Shell Oil Company), Wilmington, 98,500 bbl/d (15,660 m^3/d)
- Wilmington Refinery (Valero), Wilmington, 149,000 bbl/d (23,700 m^3/d).

Colorado:

- Commerce City Refinery (Suncor Energy (U.S.A.) Inc.), Commerce City, 100,000 bbl/d (16,000 m^3/d)

Delaware:

- Delaware City Refinery (idle in 2010; purchased by PBF Energy Partners from Valero in April 2010; currently up and processing crude oil).

Georgia

- Savannah Refinery (NuStar Energy), Savannah (Asphalt Refinery) 28,000 bpd
- Douglasville Refinery (Young Refining), Douglasville — shutdown 07/04

Hawaii:

- Kapolei Refinery (Tesoro), Kapolei 93,500 bbl/d (14,870 m^3/d)
- Hawaii Refinery (Chevron), Kapolei 54,000 bbl/d (8,600 m^3/d).

Illinois:

- Lemont Refinery (Citgo), Lemont 160,000 bbl/d (25,000 m^3/d)
- Joliet Refinery (ExxonMobil), Joliet 238,000 bbl/d (37,800 m^3/d)

- Robinson Refinery (Marathon Petroleum Company), Robinson 215,000 bbl/d (34,200 m^3/d)
- Wood River Refinery (ConocoPhillips), Wood River 306,000 bbl/d (48,700 m^3/d)

Indiana:

- Whiting Refinery (BP), Whiting 405,000 bbl/d (64,400 m^3/d)
- Mount Vernon Refinery (Countrymark Co-op), Mount Vernon 23,000 bbl/d (3,700 m^3/d)

Kansas:

- Coffeyville Refinery (Coffeyville Resources LLC), Coffeyville 112,000 bbl/d (17,800 m^3/d)
- El Dorado Refinery (Frontier Oil), El Dorado 120,000 bbl/d (19,000 m^3/d)
- McPherson Refinery (NCRA), McPherson 81,200 bbl/d (12,910 m^3/d)

Kentucky:

- Catlettsburg Refinery (Marathon Petroleum Company), Catlettsburg 222,000 bbl/d (35,300 m^3/d)
- HDG International Group Refinery, Perry 195,500 bbl/d (31,080 m^3/d)
- Somerset Refinery, Somerset 5,500 bbl/d (870 m^3/d).

Louisiana:

- Alliance Refinery (ConocoPhillips), Belle Chasse 247,000
- Baton Rouge Refinery (ExxonMobil), Baton Rouge 503,000 bbl/d (80,000 m^3/d)
- Chalmette Refinery (Chalmette Refining LLC, joint venture of ExxonMobil and PDVSA), Chalmette 193,000 bbl/d (30,700 m^3/d)
- Convent Refinery (Motiva Enterprises), Convent 255,000 bbl/d (40,500 m^3/d)
- Cotton Valley Refinery (Calumet Lubricants), Cotton Valley 13,000 bbl/d (2,100 m^3/d)
- Garyville Refinery (Marathon Petroleum Company), near Garyville 436,000 bbl/d (69,300 m^3/d)
- Krotz Springs Refinery (Alon), Krotz Springs 85,000 bbl/d (13,500 m^3/d)

- Lake Charles Refinery (Calcasieu Refining), Lake Charles 30,000 bbl/d (4,800 m^3/d)
- Lake Charles Refinery (Citgo), Lake Charles 427,800
- Lake Charles Refinery (ConocoPhillips), Westlake 247,000
- Meraux Refinery (Murphy Oil), Meraux 125,000 bbl/d (19,900 m^3/d)
- Norco Refinery (Motiva Enterprises), Norco 242,000 bbl/d (38,500 m^3/d)
- Port Allen Refinery (Placid Refining), Port Allen 48,500 bbl/d (7,710 m^3/d)
- Princeton Refinery (Calumet Lubricants), Princeton 8,300 bbl/d (1,320 m^3/d)
- Shreveport Refinery (Calumet Lubricants), Shreveport 35,000 bbl/d (5,600 m^3/d)
- St. Charles Refinery (Valero), Norco 260,000 bbl/d (41,000 m^3/d).

Michigan:

- Detroit Refinery (Marathon Petroleum Company), Detroit 180,000 bbl/d (29,000 m^3/d).

Minnesota:

- Pine Bend Refinery (Flint Hills Resources), Rosemount 320,000 bbl/d (51,000 m^3/d)
- St. Paul Park Refinery (Northern Tier Energy), St. Paul Park 70,000 bbl/d (11,000 m^3/d).

Mississippi:

- Lumberton Refinery (Hunt Southland Refining), Lumberton 5,800 bbl/d (920 m^3/d)
- Pascagoula Refinery (Chevron), Pascagoula 325,000 bbl/d (51,700 m^3/d)
- Vicksburg Refinery (Ergon), Vicksburg 23,000 bbl/d (3,700 m^3/d)
- Rogerslacy Refinery (Hunt Southland Refining), Sandersville 11,000 bbl/d (1,700 m^3/d)
- Greenville Refinery, Scott Petroleum, Biodiesel Oil Refinery

Montana:

- Billings Refinery (ConocoPhillips), Billings 58,000 bbl/d (9,200 m^3/d)

- Billings Refinery (ExxonMobil), Billings 60,000 bbl/d (9,500 m^3/d)
- Montana Refining Company (Connacher Oil & Gas Limited), Great Falls 9,500 bbl/d (1,510 m^3/d)
- Laurel Refinery (Cenex), Laurel 55,000 bbl/d (8,700 m^3/d)

Nevada:

- Eagle Springs Refinery (Foreland Refining), Currant 1,700 bbl/d (270 m^3/d).

New Jersey:

- Bayway Refinery (ConocoPhillips), Linden 230,000 bbl/d (37,000 m^3/d)
- Eagle Point Refinery (Sunoco), Westville closed 2010 145,000 bbl/d (23,100 m^3/d)
- Paulsboro Asphalt Refinery (NuStar), Paulsboro 51,000 bbl/d (8,100 m^3/d)
- Paulsboro Refinery (PBF Energy Corporation), Paulsboro 160,000 bbl/d (25,000 m^3/d)
- Perth Amboy Refinery (Chevron), Perth Amboy 80,000 bbl/d (13,000 m^3/d)
- Port Reading Refinery (Hess), Port Reading 62,000 bbl/d (9,900 m^3/d).

New Mexico:

- Artesia Refinery (Holly Corporation via Navajo Refining), Artesia 100,000 bbl/d (16,000 m^3/d)
- Bloomfield Refinery (Western Refining), Bloomfield 16,800 bbl/d (2,670 m^3/d)
- Ciniza Refinery (Western Refining), Gallup 26,000 bbl/d (4,100 m^3/d)
- Lovington Refinery (Holly Corporation), Lovington.

North Dakota:

- Mandan Refinery (Tesoro), Mandan 60,000 bbl/d (9,500 m^3/d)

Ohio:

- Canton Refinery (Marathon Petroleum Company), Canton 73,000 bbl/d (11,600 m^3/d)
- Lima Refinery (Husky Energy), Lima 158,400 bbl/d (25,180 m^3/d)
- Toledo Refinery (BP/Husky Oil), Toledo 160,000 bbl/d (25,000 m^3/d)

- Toledo Refinery (PBF Energy), Toledo 160,000 bbl/d (25,000 m^3/d).

Oklahoma:

- Ardmore Refinery (Valero), Ardmore 74,700 bbl/d (11,880 m^3/d)
- Ponca City Refinery (ConocoPhillips), Ponca City 194,000 bbl/d (30,800 m^3/d)
- Tulsa Refinery (Sinclair Oil), Tulsa 70,300 bbl/d (11,180 m^3/d)
- Tulsa Refinery (Holly Corporation), Tulsa 83,200 bbl/d (13,230 m^3/d)
- Wynnewood Refinery, Wynnewood 71,700 bbl/d (11,400 m^3/d)
- Ventura Refining and Transmission, Thomas 14,000 bbl/d (2,200 m^3/d).

Pennsylvania:

- Bradford Refinery (American Refining Group), Bradford 10,000 bbl/d (1,600 m^3/d)
- Marcus Hook Refinery (Sunoco), Marcus Hook 175,000 bbl/d (27,800 m^3/d)
- Philadelphia Refinery (Sunoco), Philadelphia 335,000 bbl/d (53,300 m^3/d)
- Penreco (Calumet), Karns City
- Trainer Refinery (ConocoPhillips), Trainer 185,000 bbl/d (29,400 m^3/d)
- Warren Refinery, United Refining Company, Warren 70,000 bbl/d (11,000 m^3/d)
- Wamsutta Oil Refinery (historical), McClintocksville
- Hess oil Refinery.

Tennessee:

- Memphis Refinery (Valero), Memphis 180,000 bbl/d (29,000 m^3/d).

Texas:

- Baytown Refinery (ExxonMobil), Baytown 560,640 bbl/d (89,135 m^3/d)
- Big Spring Refinery (Alon USA), Big Spring 61,000 bbl/d (9,700 m^3/d)
- Beaumont Refinery (ExxonMobil), Beaumont 348,500 bbl/d (55,410 m^3/d)

- Borger Refinery (ConocoPhillips/Cenovus), Borger 146,000 bbl/d (23,200 m^3/d)
- Corpus Christi Complex (Flint Hills Resources), Corpus Christi 288,000 bbl/d (45,800 m^3/d)
- Corpus Christi Refinery (Citgo), Corpus Christi 156,000 bbl/d (24,800 m^3/d)
- Corpus Christi West Refinery (Valero), Corpus Christi 142,000 bbl/d (22,600 m^3/d)
- Corpus Christi East Refinery (Valero), Corpus Christi 115,000 bbl/d (18,300 m^3/d)
- Deer Park Refinery (Shell Oil Company), Deer Park 333,700 bbl/d (53,050 m^3/d)
- El Paso Refinery (Western Refining), El Paso 120,000 bbl/d (19,000 m^3/d)
- Houston Refinery (Lyondell), Houston 270,200 bbl/d (42,960 m^3/d)
- Houston Refinery (Valero), Houston 83,000 bbl/d (13,200 m^3/d)
- Independent Refinery (Stratnor), Houston 100,000 bbl/d (16,000 m^3/d)
- McKee Refinery (Valero), Sunray 158,300 bbl/d (25,170 m^3/d)
- Pasadena Refinery (Petrobras), Pasadena 100,000 bbl/d (16,000 m^3/d)
- Port Arthur Refinery (Total), Port Arthur 174,000 bbl/d (27,700 m^3/d)
- Port Arthur Refinery (Motiva Enterprises), Port Arthur 285,000 bbl/d (45,300 m^3/d).
- Port Arthur Refinery (Valero), Port Arthur 325,000 bbl/d (51,700 m^3/d)
- Penreco (Calumet), Houston
- San Antonio Refinery (NuStar Energy), San Antonio 10,300 bbl/d (1,640 m^3/d)
- Sweeny Refinery (ConocoPhillips), Sweeny 229,000 bbl/d (36,400 m^3/d)
- Texas City Refinery (BP), Texas City 460,000 bbl/d (73,000 m^3/d)
- Texas City Refinery (Marathon Petroleum Company), Texas City 72,000 bbl/d (11,400 m^3/d)

- Texas City Refinery (Valero), Texas City 210,000 bbl/d (33,000 m^3/d)
- Three Rivers Refinery (Valero), Three Rivers 90,000 bbl/d (14,000 m^3/d)
- Tyler Refinery (Delek Refining Ltd.), Tyler 62,000 bbl/d (9,900 m^3/d).

Utah:

- North Salt Lake Refinery (Big West Oil), North Salt Lake 35,000 bbl/d (5,600 m^3/d)
- Salt Lake City Refinery (Chevron), Salt Lake City 45,000 bbl/d (7,200 m^3/d)
- Salt Lake City Refinery (Tesoro), Salt Lake City 58,000 bbl/d (9,200 m^3/d)
- Woods Cross Refinery (Holly Corporation), Woods Cross 26,000 bbl/d (4,100 m^3/d)
- Woods Cross Refinery (Silver Eagle Refining), Woods Cross 10,200 bbl/d (1,620 m^3/d).

Virginia:

- Yorktown Refinery (Western Refining), Yorktown 58,600 bbl/d (9,320 m^3/d)

Washington:

- Tesoro Anacortes Refinery (Tesoro), Anacortes 108,000 bbl/d (17,200 m^3/d)
- Shell Anacortes Refinery (Shell Oil Company), Anacortes 145,000 bbl/d (23,100 m^3/d)
- Cherry Point Refinery (BP), Blaine 225,000 bbl/d (35,800 m^3/d)
- ConocoPhillips Ferndale Refinery (ConocoPhillips), Ferndale 105,000 bbl/d (16,700 m^3/d)
- Tacoma Refinery (U.S. Oil and Refining), Tacoma 35,000 bbl/d (5,600 m^3/d).

West Virginia:

- Newell Refinery (Ergon), Newell 19,400 bbl/d (3,080 m^3/d).

Wisconsin:

- Superior Refinery (Murphy Oil), Superior 33,000 bbl/d (5,200 m^3/d).

Wyoming:

- Cheyenne Refinery (Frontier Oil), Cheyenne 52,000 bbl/d (8,300 m^3/d)
- Evanston Refinery (Silver Eagle Refining), Evanston 3,000 bbl/d (480 m^3/d)
- Evansville Refinery (Little America Refining), Evansville 24,500 bbl/d (3,900 m^3/d)
- Newcastle Refinery (Wyoming Refining), Newcastle 12,500 bbl/d (1,990 m^3/d)
- Sinclair Refinery (Sinclair Oil), Sinclair 66,000 bbl/d (10,500 m^3/d)

US Virgin Islands:

- St Croix Refinery (HOVENSA) 494,000 bpd.

New South Wales:

- Kurnell Refinery, (Caltex), 124,500 bbl/d (19,790 m^3/d), Botany Bay
- Clyde Refinery, (Royal Dutch Shell), 100,000 bbl/d (16,000 m^3/d), Clyde.

Victoria:

- Geelong Refinery, (Royal Dutch Shell), 130,000 bbl/d (21,000 m^3/d), Geelong
- Altona Refinery, (ExxonMobil), ~75,000 bpd, Altona North (refinery reduced from 2 trains to 1 train between 2000–2004)

Queensland:

- Bulwer Island Refinery, (BP), 90,000 bbl/d (14,000 m^3/d), Bulwer Island
- Lytton Refinery, (Caltex), 104,000 bbl/d (16,500 m^3/d), Lytton

South Australia:

- Port Stanvac Refinery, (ExxonMobil), 100,000 bbl/d (16,000 m^3/d), Lonsdale (mothballed since 2003 - 239 ha site to be cleaned up and redeveloped for housing)

Western Australia:

- Kwinana Refinery, (BP), 138,000 bbl/d (21,900 m^3/d), Kwinana

New Zealand:

- Marsden Point Oil Refinery (NZRC), 96,000 b.

Papua New Guinea:

- InterOil Refinery (InterOil), 32,500 hi.

South America

Argentina:

- La Plata Refinery (Repsol YPF) 189,000 bpd
- Buenos Aires Refinery (Royal Dutch Shell) 110,000 bpd
- Luján de Cuyo Refinery (Repsol YPF) 105,500 bpd
- Esso Campana Refinery (ExxonMobil) 84,500 bpd
- San Lorenzo Refinery (Refisan S.A.) 38,000 bpd (start-up 1938)
- Plaza Huincul Refinery (Repsol YPF) 37,190 bpd (start-up 1919)
- Campo Duran Refinery (Refinor) 32,000 bpd
- Bahia Blanca Refinery (Petrobras) 28,975 bpd.

Bolivia:

- Gualberto Villarael Cochabamba Refinery (YPFB) 40,000 bpd
- Guillermo Elder Bell Santa Cruz Refinery (YPFB) 20,000 bpd
- Carlos Montenegro Sucre Refinery (Refisur SA) 3,000 bpd
- Reficruz 2,000 bbl/d (320 m^3/d)
- Refineria Oro Negro SA 2,000 bbl/d (320 m^3/d).

Brazil:

- REFAP (Petrobras), Canoas 189,000 bbl/d (30,000 m^3/d)
- RECAP (Petrobras), Mauá 53,000 bbl/d (8,400 m^3/d)
- REPLAN (Petrobras), Paulinia 365,000 bbl/d (58,000 m^3/d)
- REVAP (Petrobras), São José dos Campos 251,000 bbl/d (39,900 m^3/d)
- RPBC (Petrobras), Cubatão 170,000 bbl/d (27,000 m^3/d)
- REDUC (Petrobras), Duque de Caxias 242,000 bbl/d (38,500 m^3/d)
- REMAN (Petrobras), Manaus 46,000 bbl/d (7,300 m^3/d)
- Lubnor (Petrobras), Fortaleza 6,000 bbl/d (950 m^3/d)
- REGAP (Petrobras), Betim 151,000 bbl/d (24,000 m^3/d)
- REPAR (Petrobras), Araucária 189,000 bbl/d (30,000 m^3/d)
- RLAM (Petrobras), São Francisco do Conde 323,000 bbl/d (51,400 m^3/d)

- Refinaria Ipiranga (Refinaria Riograndense), Pelotas 12,500 bbl/d (1,990 m^3/d)
- Refinaria Manguinhos (Grupo Peixoto de Castro and Repsol YPF), Rio de Janeiro 14,000 bbl/d (2,200 m^3/d).

Chile:

- BioBio Refinery (Empresa Nacional de Petroleo), 113,000 bbl/d (18,000 m^3/d)
- Aconcagua Concon Refinery (Empresa Nacional de Petroleo), 97,650 bbl/d (15,525 m^3/d)
- Gregorio Refinery (Empresa Nacional de Petroleo), 14,750 bbl/d (2,345 m^3/d).

Colombia:

- Barrancabermeja-Santander Refinery (Ecopetrol), 252,000 bpd (start-up 1922), in process to expansion to 300,000 bpd and increase the conversion.
- Cartagena Refinery (Reficar S.A.), 80,000 bpd (start-up 1957), in process expansion to 140,000 bpd.
- Apiay Refinery (Ecopetrol), 2,250 bbl/d (358 m^3/d)
- Orito Refinery (Ecopetrol), 1,800 bbl/d (290 m^3/d)
- Tibu Refinery (Ecopetrol), 1,800 bbl/d (290 m^3/d).

Ecuador:

- Esmeraldas Refinery (Petroecuador), 110,000 bpd (start-up 1978)
- La Libertad Refinery (Petroecuador), 45,000 bbl/d (7,200 m^3/d)
- Shushufindi Refinery (Petroecuador), 20,000 bbl/d (3,200 m^3/d)

Paraguay:

- Villa Elisa Refinery (Petropar) 7,500 bpd.

Peru:

- Refineria La Pampilla Lima (Repsol YPF) 102,000 bpd
- Refineria de Talara (Petroperú) 65,000 bpd (start-up 1917) with FCC unit
- Refineria Iquitos Loreto (Petroperú) 12,000 bpd (start-up 1982)
- Refineria Conchan (Petroperú) 15,000 bpd (start-up 1961)
- Refineria Pucallpa (Maple Gas) 3,250 bpd

- Refineria El Milagro (Petroperú) 1,500 bpd (start-up 1994)
- Refineria Shiviyacu (Pluspetrol) 2,000 bpd (start-up 1950).

Surinam:

- Paramaribo (Staatsolie) 7,000 bpd.

Uruguay:

- La Teja Montevideo Refinery (ANCAP) 40,000 bpd (start-up 1937)

Venezuela:

- Paraguana Refinery Complex (CRP) (PDVSA) 956,000 bdp (Amuay-Cardon-Bajo Grande) (start-up 1997)
    - o Amuay Refinery (CRP) (PDVSA) 635,000 bpd (start-up 1950)
    - o Cardon Refinery (CRP) (PDVSA) 305,000 bpd (start-up 1949)
    - o Bajo Grande Refinery (CRP) (PDVSA) 16,000 bpd (start-up 1956)
- Puerto La Cruz Refinery (PDVSA) 200,000 bpd (start-up 1948)
- El Palito Refinery (PDVSA) 140,000 bpd (start-up 1954)
- San Roque Refinery (PDVSA) 5,200 bpd.
- Upgraders (Extra Heavy Oil Joint Ventures with PDVSA at Jose)
    - o Petrozuata (PDVSA) 140,000 bpd (start-up 2000)
    - o Operadora Cerro Negro (ExxonMobil, Aral AG, and PDVSA) 120,000 bpd (start-up 2001)
    - o Sincor (Total S.A., Statoil, and PDVSA) 180,000 bpd (start-up 2001)
    - o Ameriven (ConocoPhillips, ChevronTexaco, and PDVSA) 190,000 bpd (start-up 2004).

Oil India

Oil India (OIL) is a public sector oil and gas company in India under the administrative control of the Ministry of Petroleum and Natural Gas of the Government of India. OIL is engaged in the business of exploration, development and production of crude oil and natural gas, transportation of crude oil and production of liquid petroleum gas.

The story of Oil India Limited (OIL) traces and symbolizes the development and growth of the Indian petroleum industry. From the discovery of crude oil in the far east of India at Digboi, Assam in 1889 to its present status as a fully integrated upstream petroleum company, OIL has come far, crossing many milestones.

The Company presently produces over 3.2 MMTPA (million tons per annum) of crude oil, over 5 MMSCMD of Natural Gas and over 50,000 Tones of LPG annually. Most of this emanates from its traditionally rich oil and gas fields concentrated in the Northeastern part of India and contribute to over 65% of total Oil&Gas produced in the region.

The search for newer avenues has seen OIL spreading out its operations in onshore / offshore Orissa and Andaman, deserts of Rajasthan, plains of Uttar Pradesh, riverbeds of Brahmaputra and offshore Saurashtra.

In Rajasthan, OIL discovered gas in 1988, heavy oil / bitumen in 1991 and started production of gas in 1996. The company has accumulated over a hundred years of experience in the field of oil and gas production, since the discovery of Digboi oilfield in 1889. It is possibly the only company to do so. From well completion to wellbore servicing, installation, operation and maintenance of modern surface handling facilities, the company has the skill and expertise to manage the entire range of operations required for onshore oil and gas production.

The company has over 100,000 square kilometres of license areas for oil and gas exploration. It has emerged as a consistently profitable international company with exploration blocks as far as Libya and sub-Saharan Africa.

In recent years, OIL has stepped up E & P activities significantly including Gas monetization in the North-East India. OIL has set up the NEF (North East Frontier) project to intensify its exploration activities in the frontier areas in North East, which are logistically very difficult and geologically complex. Presently, seismic surveys are being carried out in Manbhum, Pasighat and other Trust Belt areas. The Company operates a crude oil pipeline in the North East for transportation of crude oil produced by both OIL and ONGCL in the region to feed Numaligarh, Guwahati, Bongaigaon and Barauni refineries and a branch line to feed Digboi refinery.

Oil India Pipeline History

A 1157 kilometres long fully automated telemetric pipeline with 212 kilometres of looping having a total capacity to transport over 6.0 MMTPA remains the lifeline of the Company. Commissioned in 1962, the double skinned crude oil pipeline traverses 78 river crossings including the mighty Brahmaputra River meandering through paddy fields, forests and swamps. There are 9 pumping stations, 17 Repeater stations and a terminal at Barauni. The engines that drive the giant pumps along the pipeline have crossed over two hundred thousand hours of service and established a world record of machine run - hours.

The Company is currently in the process of constructing a 660 KM long Product Pipeline from Numaligarh to Siliguri. The Pipeline is expected to be completed by mid 2007. OIL also sells its produced gas to different customers in Assam viz. BVFCL, ASEB, NEEPCO, IOC (AOD), and APL and to RSEB in Rajasthan. The company also produces Liquefied Gas (LPG) in its plant at Du.

Oil India Limited

Oil India Limited (OIL) is a premier Indian National oil company under the administrative control of Ministry of Petroleum and Natural Gas, Govt. of India. OIL is engaged in the business of Exploration, Development and Production of Crude Oil and Natural Gas, Transportation of Crude Oil and Production of LPG. The Company has over 1 lakh sq. km. of license areas. Oil India Limited is the pioneer in exploration and production of hydrocarbon in India, has been serving the nation for over four decades. Oil India is an integrated upstream petroleum company performing the following main activities:

- o Exploration for hydrocarbons.
- o Production of crude oil and natural gas.
- o Transportation of crude oil to refineries.
- o Supply of gas to consumers.
- o Extraction and bottling of LPG.

Oil India owns and operates a wide array of facilities and equipment to carry out seismic and geodetic work, 2D and 3D data acquisition, processing and analysis, onshore and offshore drilling, oil and gas field development and production, LPG production and other ancillary services to make it a fully integrated E&P company.

The Company has been steadily improving its performance year after year in the areas of production, sales, accretion to reserves, etc. The physical and financial performances in the last three years are as follows:

Oil India, which has traditionally been producing around 3 MMTPA of crude and 5 MMSCMD of natural gas, has recently embarked on an overall performance-improvement exercise. It has been successful in arresting the declining trend in production of crude oil from its ageing fields in the North-East region by application of latest technology land equipment. It has, in fact, recorded a growth of 20 per cent in the last 18 months. It is worth mentioning that over the years the Reserve Accretion Ratio for Oil India has consistently been over 1. The company is optimistic of continuing with its growth plan and expects to achieve 4.00 MMTPA of crude oil and 7 MMSCM of natural gas in the next financial year, i.e.2006-07.

Asia's first cross-country Pipeline is owned and operated by Oil India. This 1157 KM long pipeline traverses through three states, viz. Assam, West Bengal and Bihar, is the lifeline of the North-East as it carries the crude oil, which is very vital for the survival of the four Refineries in the region.

The Company is currently in the process of constructing a 660 KM long Product Pipeline from Numaligarh to Siliguri. The Pipeline, when commissioned in early 2006, is expected to solve the product evacuation problem of the modern and state-of-the-art Numaligarh Refinery which has been set up under the Assam Accord with the objective of creating all-round socio-economic development in the State.

The Company's operational areas are spread in the states of Assam, Arunachal Pradesh, Orissa, Rajasthan, Uttar Pradesh and Uttaranchal. Active participation in the NELP bids has helped the Company to acquire interests in 13 Blocks and with Operatorship in five of them. Out of the balance eight Blocks, five Blocks are in deep sea, wherein ONGCL is the operator with participating interest in the range of 15-20%. The Company is determined to aggressively bid for Blocks offered in the NELP-V Rounds which is due for closure on 31.5.2005.

Oil India has identified acquisition of overseas exploration and production ventures as an essential requirement for fast growth. The

Company already has participating interests in various overseas projects, such as, in Iran, Cote d'lvoire, and Sudan. It has recently formed a 50:50 strategic alliance with Indian Oil Corporation Limited (IOC) to jointly pursue exploration and production opportunities abroad and to achieve synergy in the downstream sector. This consortium has achieved a major breakthrough in its maiden attempt by winning a Block in the highly prospective Sirte Basin in Libya with the Operatorship for Oil India, against competitive bidding.

As a corporate citizen, Oil India has always been and is conscious of its responsibility towards Socio-Economic Development. The Organization's deep commitment towards this objective is obvious from the fact that it has adopted a separate Vision Statement for this purpose. The Company earmarks a substantial amount annually for carrying out socio-economic initiatives, which also include health care and education, in all the areas where it has operations.

5

Oil Depletion

Oil depletion occurs in the second half of the production curve of an oil well, oil field, or the average of total world oil production. The Hubbert peak theory makes predictions of production rates based on prior discovery rates and anticipated production rates. Hubbert curves predict that the production curves of non-renewing resources approximate a bell curve. Thus, when the peak of production is passed, production rates enter an exponential decline.

The American Petroleum Institute estimated in 1999 the world's oil supply would be depleted between 2062 and 2094, assuming total world oil reserves at between 1.4 and 2 trillion barrels (220 and 320 km^3) and consumption at 80 million barrels per day (13,000,000 m^3/d). In 2004, total world reserves were estimated to be 1.25 trillion barrels (199 km^3) and daily consumption was about 85 million barrels (13,500,000 m^3), shifting the estimated oil depletion year to 2057.

A study published in the journal Energy Policy by researchers from Oxford University, however, predicted demand would surpass supply by 2015 (unless constrained by strong recession pressures caused by reduced supply or government intervention).

The United States Energy Information Administration predicted in 2006 that world consumption of oil will increase to 98.3 million barrels per day (15,630,000 m^3/d) (mbd) in 2015 and 118 mbd in 2030.

With 2009 world oil consumption at 84.4 mbd, reaching the projected 2015 level of consumption would represent an average annual increase between 2009 and 2015 of 2.7% per year while EIA's own figures show declining consumption and declining supplies during the 2005-2009 period.

Resource Availability

The world's oil supply is fixed because petroleum is naturally formed far too slowly to be replaced at the rate at which it is being extracted. Over many millions of years, plankton, bacteria, and other plant and animal matter become buried in sediments on the ocean floor. When conditions are right – a lack of oxygen for decomposition, and sufficient depth and temperature of burial – these organic remains are converted into petroleum compounds, while the sediment accompanying them is converted into sandstone, siltstone, and other porous sedimentary rock. When capped by impermeable rocks such as shale, salt, or igneous intrusions, they form the petroleum reservoirs which are exploited today.

Production Decline Models

Oil production decline occurs in a predictable manner based on geological circumstances, governmental policies, and engineering practices. The shape of the decline curve varies depending upon whether one considers a well, a field, a set of fields, or the world.

Oil Well Production Decline

Oil well production curves typically end in an exponential decline. At natural rates, oil well production curves appear similar to a bell curve, a phenomenon known as the Hubbert curve. The typical decline is a rapid drop in production, and eventually a leveling off to a point at which they no longer produce profitable amounts. Such wells are referred to as marginal or stripper wells.

The shape of production curve of an oil well can be affected by a number of factors:

- Well may be restricted by choice by lack of market demand or government regulation. This flattens the peak of the curve, but will not change the well's total production significantly.
- Hydraulic fracturing (*fracing*) or acidizing may be used to cause a sharp spike in production, and may increase the recoverable reserves of a given well.
- The field may undergo a secondary or tertiary recovery project, discussed in the next section.

Oil Field Production Decline

Each individual oil well is a portion of a larger fixed area oil field. As with individual wells, discovery and production amounts of oil fields generally average to a similar bell shaped production curve. Eventually, when the field is completely drilled out, a field's production goes into a sharp decline as the average production of its wells enter decline. As this decline levels off, production can continue at relatively low rates. A number of oil fields in the U.S. have been producing for over 100 years.

Oil field production curves can be modified by a number of factors:

- Production may be restricted by market conditions or government regulation.
- A secondary recovery project, such as water or gas injection, can repressurize the field and improve the production rate temporarily. However, it will not change the total production amount over the life of the field. Eventually the field will go into a steeper than normal decline.
- the field may undergo an enhanced oil recovery project, such as drilling of wells for injection of solvents, carbon dioxide, or steam. This can be very expensive but allows more oil to be coaxed out of the rock, increasing the ultimate production of the field.

Multi-field Production Decline

Most oil is found in a small number of very large oil fields. If oil fields are discovered at a constant rate until they have all been found, the combined production of fields will yield a curve such as the one at right. Production starts off slowly, rises faster and faster, then slows down and flattens until it reaches a peak. After the production peak, production enters an exponential decline, eventually flattening out. Oil production may never actually reach zero, but eventually becomes very low. Factors which can modify this curve include:

- Inadequate demand for oil, which reduces steepness of the curve and pushes its peak into the future.
- Sharp price increases when the production peak is reached, as production fails to meet demand. If price increases cause a sharp drop in demand, a dip in the top of the curve may occur.
- Development of new drilling technology or marketing of unconventional oil can reduce the steepness of the decline as more oil is produced than initially anticipated.

United States Production Decline

Oil production in the United States has followed the theoretical Hubbert curve. U.S. oil production reached its peak in 1970 and by the mid-2000s it had fallen to 1940s levels. In 1950, the United States produced over half the world's oil, but by 2005 that proportion had dropped to about 8%. In 2005, U.S. crude oil imports were twice as high as domestic production.

The production peak in 1970 was predicted in 1956 by Hubbert. By 1972 all import quotas and controls on U.S. domestic production had been removed. Oil companies began drilling large numbers of oil wells on a nationwide scale. Despite this, and despite the quadrupling of prices during the 1973 oil crisis, the production decline has, to date, proven irreversible.

The actual U.S. production curve does deviate from Hubbert's 1956 curve in some significant ways:

- When oil surpluses created a glut on the market and low prices began causing demand and production curves to rise, regulatory agencies such as the Texas Railroad Commission stepped in to restrain production.
- The curve peaked at a sharp point rather than gradually flattening out. This occurred because as consumption began to approach production limits, oil companies drilled out all their existing fields as fast as they could, and many of those fields peaked simultaneously.
- Production fell after 1970, but started to recover and reached a lower secondary peak in 1988. This occurred because the supergiant Prudhoe Bay field in Alaska was only discovered in 1968, and the Trans-Alaska Pipeline System (TAPS) was not completed until 1977. After 1988, Alaska production peaked and total U.S. production began to decline again. By 2005, Prudhoe Bay had produced over 75% of its oil.

World Oil Production

World oil production has followed a typical Hubbert curve, rising over the past century with only a few dips. The 1970 production peak in the U.S. caused many people to begin to question when the world production peak would occur. The peak of world production is known as Peak oil. By the mid-2000s, all of the world's major oil producing countries except Saudi Arabia were producing at maximum capacity (many having peaked in production), and some experts such as Matthew Simmons were questioning whether even Saudi Arabia had any reserve capacity left.

Industry observers have pointed to the similarities between the global production curve in mid-2000s and that of the United States in the 1970s.

- The oil price increases since 2003 were preceded by a decade of production cutbacks in OPEC countries in an attempt to keep prices high despite an oil glut. This is similar to production cutbacks in Texas and other states to maintain prices despite an oil glut in the decade prior to the 1973 oil crisis.
- World oil prices reached record inflation adjusted highs beginning in 2008, but new oil did not appear on the market, as the theory of supply and demand would predict. This is reminiscent of price increases in the United States in the 1970s when U.S. oil production started to decline despite record high prices and record drilling by oil companies.
- There are serious doubts about whether OPEC countries really have the oil reserves they claim. This is similar to the illusionary oil reserves that U.S. oil companies claimed to have in the decade prior to the 1973 and 1979 oil crisis. In the 1970s, those companies were unable to produce as much oil as they had predicted, and production went down instead of up.

Implications of a World Peak

A peak in oil production could result in a worldwide oil shortage, or it could not even be noticed as demand decreases in conjunction with increased prices. While past shortages stemmed from a temporary insufficiency of supply, crossing Hubbert's Peak would mean that the production of oil would continue to decline, and that demand for these products must be reduced to meet supply. The effects of such a shortage would depend on the rate of decline and the development and adoption of effective alternatives.

If alternatives were not forthcoming, it has been speculated that the numerous products produced with oil would become scarcer, leading to at the very least lower living standards in developed and developing countries alike, and possibly in the worst case to the collapse of the entire international banking system, which could not likely sustain itself without the prospect of growth. The political situation may change dramatically, with potential wars between countries over access to dwindling supplies. Accordingly, inequalities between various countries and regions of the world may become exacerbated.

Catastrophe

Economic growth and prosperity since the industrial revolution have, in large part, been due to increased efficiencies in the use of better and higher concentrations of energy in fossil fuels. The use of fossil fuels allows humans to participate in takedown, which is the consumption of energy at a greater rate than it is being replaced. Some believe that decreasing oil production portends a drastic impact on human culture and modern technological society, which is currently heavily dependent on oil as a fuel and chemical feedstock. For example, over 90% of transportation in the United States relies on oil.

Some envisage a Malthusian catastrophe occurring as oil becomes increasingly inefficient to produce, others have learned from the examples demonstrated in mature basins and applied those operational procedures to these basins to preserve their operational tempo. Since the 1940s, agriculture has dramatically increased its productivity, due largely to the use of chemical pesticides, fertilizers, and increased mechanisation.

This process has been called the Green Revolution. The increase in food production has allowed world population to grow dramatically over the last 50 years. Pesticides rely upon oil as a critical ingredient, and fertilizers require natural gas. Farm machinery also requires oil.

Arguing that in today's world every joule one eats requires 5–15 joules to produce and deliver, some have speculated that decreasing supply of oil will cause modern industrial agriculture to collapse, leading to a drastic decline in food production, food shortages and possibly even mass starvation.

However, most or all of the uses of fossil fuels in agriculture can be replaced with alternatives. For example, by far the biggest fossil fuel input to agriculture is the use of natural gas as a hydrogen source for the Haber-Bosch fertilizer-creation process. Natural gas is used simply because it is the cheapest currently-available source of hydrogen; were that to change, other sources, such as electrolysis powered by solar energy, could be used to provide the hydrogen for creating fertilizer without relying on fossil fuels.

Oil shortages may force a move to lower input "organic agriculture" methods, which may be more labor-intensive and require a population shift from urban to rural areas, reversing the trend towards urbanisation which has predominated in industrial societies; however, some organic farmers using modern organic-farming methods have reported yields as high as those available from conventional farming, but without the use of fossil-fuel-intensive artificial fertilizers or pesticides.

Another possible effect would derive from modern transportation and housing infrastructure. A large proportion of the developed world's population live in suburbs, a type of low-density settlement designed with the automobile in mind. Commentators such as James Howard Kunstler argue that because of its reliance on the automobile, the suburb is an unsustainable living arrangement; the implications of peak oil would leave many suburb-dwellers unable to afford fuel for their cars, and force them to move to higher density, more walkable areas. In effect, surburbia would comprise the "slums of the future." A movement to deal with this problem early, called "New Urbanism," seeks to develop the suburbs into higher density neighborhoods and use high density, mixed-use forms for new building projects.

Recession

A more modest scenario, assuming a slower rate of depletion and a smooth transition to alternative energy sources could cause substantial economic hardship such as a recession or depression due to higher energy prices. Historically, there is a close correlation in the timing of oil price spikes and economic downturns. Inflation has also been linked to oil price spikes. However, economists disagree on the strength and causes of this association. Conversely, the recessions of the early 1970s and early 1980s were associated with a relatively brief period of somewhat dwindling energy availability; the possible future increase in oil prices might be much higher and last longer. See Energy crisis.

Rising Food Prices

Rising oil prices cause rising food prices in three ways. First, increased equipment fuel costs drive higher prices. Second, transportation costs increase retail prices. Third, higher oil prices are causing farmers to switch from producing food crops to producing biofuel crops. The law of supply and demand predicts that if fewer farmers are producing food the price of food will rise.

Oil Fields

This list of oil fields includes some major oil fields of the past and present. The list is incomplete; there are more than 40,000 oil and gas fields of all sizes in the world. However, 94% of known oil is concentrated in fewer than 1500 giant and major fields. The largest fields are located in Siberia and the Middle East.

Amounts given are estimated ultimate recoverable resources (proved reserves plus cumulative production), given current technology, in barrels. Oil shale reserves (perhaps 3 trillion barrels (4.8×10^{11} m^3)) and non-conventional sources are not included in this chart. Coal (which can be converted to liquid petroleum) is also not included in this chart.

Gasoline

Figure: *A jar containing gasoline*

Gasoline is a toxic translucent, petroleum-derived liquid that is primarily used as a fuel in internal combustion engines. It consists mostly of organic compounds obtained by the fractional distillation of petroleum, enhanced with a variety of additives. Some gasolines also contain ethanol as an alternative fuel. In North America, the term "gasoline" is often shortened in colloquial usage to "gas", whereas most current or former Commonwealth nations use the term "petrol". Under normal ambient conditions its material state is liquid, unlike liquefied petroleum gas or "natural gas".

Properties

Volatility: Gasoline is more volatile than diesel oil, Jet-A, or kerosene, not only because of the base constituents, but also because of additives. Volatility is often controlled by blending with butane, which boils at -0.5 °C. The volatility of gasoline is determined by the Reid vapor pressure (RVP) test. The desired volatility depends on the ambient temperature. In hot weather, gasoline components of higher molecular weight and thus lower volatility are used. In cold weather, too little volatility results in cars failing to start.

In hot weather, excessive volatility results in what is known as "vapor lock", where combustion fails to occur, because the liquid fuel has changed to a gaseous fuel in the fuel lines, rendering the fuel pump ineffective and starving the engine of fuel. This effect mainly applies to camshaft-driven (engine mounted) fuel pumps which lack a fuel return line. Vehicles with fuel injection require the fuel to be pressurized, to within a set range. Because camshaft speed is nearly zero before the engine is started, an electric pump is used. It is located in the fuel tank so the fuel may also cool the high-pressure pump. Pressure regulation is achieved by returning unused fuel to the tank. Therefore, vapor lock is almost never a problem in a vehicle with fuel injection.

In the United States, volatility is regulated in large cities to reduce the emission of unburned hydrocarbons by the use of so-called reformulated gasoline that is less prone to evaporation. In Australia, summer petrol volatility limits are set by state governments and vary among states. Most countries simply have a summer, winter, and perhaps intermediate limit.

Volatility standards may be relaxed (allowing more gasoline components into the atmosphere) during gasoline shortages. For example, on 31 August 2005, in response to Hurricane Katrina, the United States permitted the sale of nonreformulated gasoline in some urban areas, effectively permitting an early switch from summer to winter-grade gasoline. As mandated by EPA administrator Stephen L. Johnson, this "fuel waiver" was made effective until 15 September 2005.

Modern automobiles are also equipped with an evaporative emissions control system (called an EVAP system in automotive jargon), which collects evaporated fuel from the fuel tank in a charcoal-filled canister while the engine is stopped, and then releases the collected vapors into the engine intake for burning when the engine is running (usually only after it has reached normal operating temperature).

The evaporative emissions control system also includes a sealed gas cap to prevent vapors from escaping via the fuel filler tube. Modern vehicles with OBD-II emissions control systems will illuminate the malfunction indicator light (MIL), "check engine" or "Service Engine Soon" light if the leak detection pump (LDP) detects a leak in the EVAP system. If the electronic control unit (ECU) or powertrain control module (PCM) detects a leak, it will store an OBD-II code representing either a small or large leak, thus illuminating the MIL to indicate a failure. Some vehicles can detect whether the gas cap is incorrectly fitted, and will indicate this by illuminating a gas cap symbol on the dash.

Octane Rating

Internal combustion engines are designed to burn gasoline in a controlled process called deflagration. But in some cases, gasoline can also combust abnormally by detonation, which wastes energy and can damage the engine. One way to reduce detonation is to increase the gasoline's resistance to autoignition, which is expressed by its octane rating. Octane rating is measured relative to a mixture of 2,2,4-trimethylpentane (an isomer of octane) and n-heptane. There are different conventions for expressing octane ratings, so a fuel may have several different octane ratings based on the measure used. Research octane number (RON) for gasoline varies with country.

In Finland, Sweden, and Norway, 95 RON is the standard for regular unleaded petrol and 98 RON is also available as a more expensive option. In the UK, ordinary regular unleaded petrol is 91 RON (not commonly available), premium unleaded petrol is always 95 RON, and super unleaded is usually 97-98 RON.

However, both Shell and BP produce fuel at 102 RON for cars with high-performance engines, and the supermarket chain Tesco began in 2006 to sell super unleaded petrol rated at 99 RON. In the US, octane ratings in unleaded fuels can vary between 86 and 87 AKI (91-92 RON) for regular, through 89-90 AKI (94-95 RON) for mid-grade (European premium), up to 90-94 AKI (95-99 RON) for premium (European super).

The octane rating became important as the military sought higher output for aircraft engines in the late 1930s and the 1940s. A higher octane rating allows a higher compression ratio, and thus higher temperatures and pressures, which translate to higher power output. Some scientists even predicted that nation with a good supply of high octane gasoline would have the advantage in air power

Stability

Good quality gasoline should be stable almost indefinitely if stored properly. Such storage should be in an airtight container, to prevent oxidation or water vapors mixing, and at a stable cool temperature, to reduce the chance of the container leaking. When gasoline is not stored correctly, gums and solids may accumulate resulting in "stale fuel". The presence of these degradation products in fuel tank, lines, and carburetor or fuel injection components, make it harder to start the engine. Upon the resumption of regular vehicle usage, though, the buildups should eventually be cleaned up by the flow of fresh petrol. Fuel stabilizers can be used to extend the life of the fuel that is not or cannot be stored properly. Fuel stabilizer is commonly used for small engines, such as lawnmower and tractor engines, to promote quicker and more reliable starting. Users have been advised to keep gasoline containers and tanks more than half full and properly capped to reduce air exposure, to avoid storage at high temperatures, to run an engine for ten minutes to circulate the stabilizer through all components prior to storage, and to run the engine at intervals to purge stale fuel from the carburetor.

Energy Content (high and low heating value)

Energy is obtained from the combustion of gasoline, the conversion of a hydrocarbon to carbon dioxide and water. The combustion of octane follows this reaction:

$$2\ C_8H_{18} + 25\ O_2 \rightarrow 16\ CO_2 + 18\ H_2O$$

Combustion of one US gallon of gasoline produces about 19.4 pounds (8.8 kg) of carbon dioxide (converts to 2.33 kg/litre), a greenhouse gas.

Gasoline contains about 35 MJ/L (9.7 kW·h/L, 132 MJ/US gal, 36.6 kWh/US gal) (higher heating value) or 13 kWh/kg. Gasoline blends differ, and therefore actual energy content varies according to the season to season and producer by up to 4% more or less than the average, according to the US EPA. On average, about 19.5 US gallons (16.2 imp gal; 74 L) of gasoline are available from a 42-US-gallon (35 imp gal; 160 L) barrel of crude oil (about 46% by volume), varying due to quality of crude and grade of gasoline. The remaining residue comes off as products ranging from tar to naptha.

A high octane fuel, such as liquefied petroleum gas (LPG), has a lower energy content than lower octane gasoline, resulting in an overall lower power output at the regular compression ratio of an engine run at on gasoline. However, with an engine tuned to the use of LPG (i.e. via higher compression ratios, such as 12:1 instead of 8:1), this lower power output can be overcome.

This is because higher-octane fuels allow for a higher compression ratio hence a higher cylinder temperature, which improves efficiency. Also, increased mechanical efficiency is created by a higher compression ratio through the concommitant higher expansion ratio on the power stroke, which is by far the greater effect. The higher expansion ratio extracts more work from the high-pressure gas created by the combustion process.

The applicable formula is PV=nRT. An Atkinson cycle engine uses the timing of the valve events to produce the benefits of a high expansion ratio without the disadvantages, chiefly detonation, of a high compression ratio. A high expansion ratio is also one of the two key reasons for the efficiency of Diesel engines, along with the elimination of pumping losses due to throtttling of the intake air flow. A high compression ratio can be viewed as a necessary evil to have a high expansion ratio.

The lower energy content (per litre) of LPG in comparison to gasoline is due mainly to its lower density. Energy content per kilogram is higher than for gasoline (higher hydrogen to carbon ratio). The weight-density of gasoline is about 740 kg/m^3 (6.175 lb/US gal; 7.416 lb/imp gal).

Density

The specific gravity (or relative density) of gasoline ranges from 0.71–0.77 (719.7 kg/m^3 ; 0.026 lb/in^3; 6.073 lb/US gal; 7.29 lb/imp gal), higher densities having a greater volume of aromatics. Gasoline floats on water; water cannot generally be used to extinguish a gasoline fire, unless used in a fine mist.

Chemical Analysis and Production

Gasoline is produced in oil refineries. Material that is separated from crude oil via distillation, called virgin or straight-run gasoline, does not meet the required specifications for modern engines (in particular octane rating), but will form part of the blend.

Figure 2: *Some of the main components of gasoline: isooctane, butane, an aromatic compound, and the octane enhancer MTBE.*

The bulk of a typical gasoline consists of hydrocarbons with between four and 12 carbon atoms per molecule (commonly referred to as C4-C12).

Figure: *Refinery in Minatitlán, Mexico*

The various refinery streams blended to make gasoline have different characteristics. Some important streams are:

- straight-run gasoline is distilled directly from crude oil. Once the leading source of fuel, its low octane rating required lead additives. It is in low aromatics (depending on the grade of crude oil), containing some naphthenes (cycloalkanes) and no olefins. About 0-20% of gasoline is derived from this material, in part because the supply of this fraction is insufficient and its RON is too low.

- reformate, produced in a catalytic reformer with a high octane rating and high aromatic content, and very low olefins (alkenes). Most of the benzene, toluene, and xylene (the so-called BTX) are more valuable as chemical feedstocks and are thus removed to some extent.
- cat cracked gasoline or cat cracked naphtha, produced from a catalytic cracker, with a moderate octane rating, high olefins (alkene) content, and moderate aromatics level.
- hydrocrackate (heavy, mid, and light) produced from a hydrocracker, with medium to low octane rating and moderate aromatic levels.
- alkylate is produced in an alkylation unit, involving the addition of isobutane to alkenes giving branched chains but low aromatics.
- isomerate is obtained by isomerizing low octane straight run gasoline to iso-parafins (like isooctane).

Figure: *A pumpjack in the United States*

The terms above are the jargon used in the oil industry but terminology varies.

Overall, a typical gasoline is predominantly a mixture of paraffins (alkanes), naphthenes (cycloalkanes), and olefins (alkenes). The actual ratio depends on:

- the oil refinery that makes the gasoline, as not all refineries have the same set of processing units;
- crude oil feed used by the refinery;
- the grade of gasoline, in particular, the octane rating.

Currently, many countries set limits on gasoline aromatics in general, benzene in particular, and olefin (alkene) content. Such regulations led to increasing preference for high octane pure paraffin (alkane) components, such as alkylate, and is forcing refineries to add processing units to reduce benzene content.

Gasoline can also contain other organic compounds, such as organic ethers (deliberately added), plus small levels of contaminants, in particular organosulfur compounds, but these are usually removed at the refinery.

Additives

Antiknock Additives: Most countries have phased out leaded fuel. Different additives have replaced the lead compounds. The most popular additives include aromatic hydrocarbons, ethers and alcohol (usually ethanol or methanol).

Tetraethyl Lead

Gasoline, when used in high-compression internal combustion engines, has a tendency to autoignite (*detonate*) causing damaging "engine knocking" (also called "pinging" or "pinking") noise. Early research into this effect was led by A.H. Gibson and Harry Ricardo in England and Thomas Midgley and Thomas Boyd in the United States. The discovery that lead additives modified this behavior led to the widespread adoption of their use in the 1920s, and therefore more powerful, higher compression engines. The most popular additive was tetra-ethyl lead. With the discovery of the extent of environmental and health damage caused by the lead, however, and the incompatibility of lead with catalytic converters found on virtually all newly sold US automobiles since 1975, this practice began to wane (encouraged by many governments introducing differential tax rates) in the 1980s.

In the US, where lead had been blended with gasoline (primarily to boost octane levels) since the early 1920s, standards to phase out leaded gasoline were first implemented in 1973 - due in great part to studies conducted by Philip J. Landrigan. In 1995, leaded fuel accounted for only 0.6% of total gasoline sales and less than 2000 short tons (1814 t) of lead per year.

From 1 January 1996, the Clean Air Act banned the sale of leaded fuel for use in on-road vehicles. Possession and use of leaded gasoline in a regular on-road vehicle now carries a maximum $10,000 fine in the US. However, fuel containing lead may continue to be sold for off-road uses, including aircraft, racing cars, farm equipment, and marine engines. Similar bans in other countries have resulted in lowering levels of lead in people's bloodstreams.

MMT

Methylcyclopentadienyl manganese tricarbonyl (MMT) has been used for many years in Canada and recently in Australia to boost octane. It also helps old cars designed for leaded fuel run on unleaded fuel without need for additives to prevent valve problems.

US Federal sources state MMT is suspected to be a powerful neurotoxin and respiratory toxin, and a large Canadian study concluded that MMT impairs the effectiveness of automobile emission controls and increases pollution from motor vehicles.

In 1977, use of MMT was banned in the US by the Clean Air Act until the Ethyl Corporation could prove the additive would not lead to failure of new car emission-control systems. As a result of this ruling, the Ethyl Corporation began a legal battle with the EPA, presenting evidence that MMT was harmless to automobile emissions-control systems. In 1995, the US Court of Appeals ruled that the EPA had exceeded its authority, and MMT became a legal fuel additive in the US. MMT is now manufactured by the Afton Chemical Corporation division of Newmarket Corporation.

Fuel Stabilizers (antioxidants and metal deactivator)

Gummy, sticky resin deposits result from oxidative degradation of gasoline upon long term storage. They arise from the reaction of oxidation of alkenes. Improvements in refinery techniques have generally reduced the reliance on the catalytically or thermally cracked gasolines that are most susceptible to oxidation.

This degradation can be prevented through the addition of 5-100 ppm of antioxidants, such as phenylenediamines and other amines. Hydrocarbons with a bromine number of 10 or above can be protected with the combination of unhindered or partially hindered phenols and oil soluble strong amine bases, such as hindered phenols. "Stale" gasoline can be detected by a colorimetric enzymatic test for organic peroxides produced by oxidation of the gasoline.

Gasolines are also treated with metal deactivators, which are compounds that sequester (deactivate) metal salts that otherwise accelerate the formation of gummy residues. The metal impurities might arise from the engine itself or as contaminants in the fuel.

Detergents

Gasoline, as delivered at the pump, also contains additives to reduce internal engine carbon buildups, improve combustion, and to allow easier starting in cold climates. High levels of detergent can be found in Top Tier Detergent Gasolines. These gasolines exceed the U.S. EPA's minimum requirement for detergent content. The specification for Top Tier Detergent Gasolines was developed by four automakers: GM, Honda, Toyota and BMW. According to the bulletin, the minimal EPA requirement is not sufficient to keep engines clean. Typical detergents include alkylamines and alkyl phosphates at the level of 50-100 ppm.

Ethanol

European Union: In the EU, 5% ethanol can be added within the common gasoline spec (EN 228). Discussions are ongoing to allow 10% blending of ethanol (available in French gas stations). Most gasoline sold in Sweden has 5-15% ethanol added.

Brazil: In Brazil, the Brazilian National Agency of Petroleum, Natural Gas and Biofuels (ANP) requires gasoline for automobile use to have from 18 to 25% of ethanol added to its composition.

Australia: Legislation limits ethanol use to 10% of gasoline in Australia. It is commonly called E10 by major brands, and is less expensive than regular unleaded petrol. It is also required for retailers to label fuels containing ethanol on the dispenser.

United States: In most states, ethanol is added by law to a minimum level which is currently 5.9%. Most fuel pumps display a sticker stating the fuel may contain up to 10% ethanol, an intentional disparity which allows the minimum level to be raised over time without requiring modification of the literature/labelling. Until late 2010, fuels retailers were only authorized to sell fuel containing up to 10 percent ethanol (E10), and most vehicle warranties (except for flexible fuel vehicles) authorize fuels that contain no more than 10 percent ethanol. In parts of the United States, ethanol is sometimes added to gasoline without an indication that it is a component in some states.

Dye

In Australia, petrol tends to be dyed a light shade of purple. In the United States, the most commonly used aircraft gasoline, avgas, or aviation gas, is known as 100LL (100 octane, low lead) and is dyed blue. Red dye has been used for identifying untaxed (off highway use) agricultural diesel. The UK uses red dye to differentiate between regular diesel fuel, (often referred to as DERV from *Diesel-Engined Road Vehicle*), which is undyed, and diesel intended for agricultural and construction vehicles like excavators and bulldozers. Red diesel is still occasionally used on HGVs which use a separate engine to power a loader crane. This is a declining practice, however, as many loader cranes are powered directly by the tractor unit. In India, where leaded fuels are mainstream, petrol is dyed red whereas in South Africa unleaded fuel is dyed green and lead-replacement fuel is dyed red.

Oxygenate Blending

Oxygenate blending adds oxygen-bearing compounds such as MTBE, ETBE and ethanol. The presence of these oxygenates reduces the amount of carbon monoxide and unburned fuel in the exhaust gas. In many areas throughout the US, oxygenate blending is mandated by EPA regulations to reduce smog and other airborne pollutants. For example, in Southern California, fuel must contain 2% oxygen by weight, resulting in a mixture of 5.6% ethanol in gasoline. The resulting fuel is often known as reformulated gasoline (RFG) or oxygenated gasoline, or in the case of California, California reformulated gasoline. The federal requirement that RFG contain oxygen was dropped on 6 May 2006 because the industry had developed VOC-controlled RFG that did not need additional oxygen.

MTBE use is being phased out in some states due to issues with contamination of ground water. In some places, such as California, it is already banned. Ethanol and, to a lesser extent, the ethanol-derived ETBE are common replacements. Since most ethanol is derived from biomass, such as corn, sugar cane or grain, it is referred to as bioethanol. A common ethanol-gasoline mix of 10% ethanol mixed with gasoline is called gasohol or E10, and an ethanol-gasoline mix of 85% ethanol mixed with gasoline is called E85. The most extensive use of ethanol takes place in Brazil, where the ethanol is derived from sugarcane. In 2004, over 3.4 billion US gallons (2.8 billion imp gal/ 13 million m^3) of ethanol was produced in the United States for fuel use, mostly from corn, and E85 is slowly becoming available in much of the United States, though many of the relatively few stations vending E85 are not open to the general public.

The use of bioethanol, either directly or indirectly by conversion of such ethanol to bio-ETBE, is encouraged by the European Union Directive on the Promotion of the use of biofuels and other renewable fuels for transport. Since producing bioethanol from fermented sugars and starches involves distillation, though, ordinary people in much of Europe cannot legally ferment and distill their own bioethanol at present (unlike in the US, where getting a BATF distillation permit has been easy since the 1973 oil crisis).

Safety

Environmental Considerations: Hydrocarbons are hazardous substances and are regulated in the United States by the Occupational Safety and Health Administration. The material safety data sheet for unleaded gasoline shows at least 15 hazardous chemicals occurring in various amounts, including benzene (up to 5% by volume), toluene (up to 35% by volume), naphthalene (up to 1% by volume), trimethylbenzene (up to 7% by volume), methyl *tert*-butyl ether (MTBE) (up to 18% by volume, in some states) and about ten others. Benzene and many antiknocking additives are carcinogenic. The chief risks of such leaks come not from vehicles, but from gasoline delivery truck accidents and leaks from storage tanks. Because of this risk, most (underground) storage tanks now have extensive measures in place to detect and prevent any such leaks, such as sacrificial anodes.

The main concern with gasoline on the environment, aside from the complications of its extraction and refining, are the potential effect on the climate. Unburnt gasoline and evaporation from the tank, when in the atmosphere, react in sunlight to produce photochemical smog. Addition of ethanol increases the volatility of gasoline, potentially worsening the problem.

Inhalation

Hydrocarbons including exhibit low acute toxicities, with LD50 of 700 – 2700 mg/kg for simple aromatic compounds. Petrol sniffing is a common intoxicant that has become epidemic in some poorer communities and indigenous groups in Australia, Canada, New Zealand, some Pacific Islands, and the US. In response, Opal fuel has been developed by the BP Kwinana Refinery in Australia, and contains only 5% aromatics (unlike the usual 25%) which weakens the effects of inhalation.

Flammability

Like other alkanes, gasoline burns in a limited range of its vapor phase and, coupled with its volatility, this makes leaks highly dangerous when sources of ignition are present. Gasoline has a lower explosion limit of 1.4% by volume and an upper explosion limit of 7.6%. If the concentration is below 1.4% the air-gasoline mixture is too lean and will not ignite. If the concentration is above 7.6% the mixture is too rich and also will not ignite. However, gasoline vapor rapidly mixes and spreads with air, making unconstrained gasoline quickly flammable. Many accidents involve gasoline being used in an attempt to light bonfires; rather than helping the material on the bonfire to burn, some of the gasoline vaporises quickly after being poured and mixes with the surrounding air, so when the fire is lit a moment later, the vapor surrounding the bonfire instantly ignites in a large fireball, engulfing the unwary user. The vapor is also heavier than air and tends to collect in garage inspection pits.

Usage and Pricing

The US accounts for about 44% of the world's gasoline consumption. In 2003 The US consumed 476.474 gigalitres (1.25871×10^{11} US gal; 1.04810×10^{11} imp gal), which equates to 1.3 gigalitres of gasoline each day (about 360 million US or 300 million imperial gallons). The US used about 510 billion litres (138 billion US gal/115 billion imp gal) of gasoline in 2006, of which 5.6% was mid-grade and 9.5% was premium grade.

Western countries have among the highest usage rates per person.

Europe

Unlike the US, countries in Europe impose a substantial taxes on fuels such as gasoline. For example, price for gasoline in Europe is more than twice that in the US.

Table: *Pump price (in Euro/liter) 2004 to 2011 lead-free 95 Octane gasoline in selected European countries. To convert prices for Euro/liter to US$/gal, multiply by 5.7 (assuming US$1.5 = 1 Euro).*

Country	*Dec. 2004*	*May 2005*	*July 2007*	*April 2008*	*Jan 2009*	*Mar 2010*	*Feb 2011*
Germany	1.19	1.18	1.37	1.43	1.09	1.35	1.50
France	1.05	1.15	1.31	1.38	1.07	1.35	1.53
Italy	1.10	1.23	1.35	1.39	1.10	1.34	1.46
Netherlands	1.26	1.33	1.51	1.56	1,25	1.54	1.66
Poland	0.80	0.92	1.15	1.23	0.82	1.12	1.26
Switzerland	0.92	0.98	1.06	1.14	0.88	1.12	1.29
Hungary	1.00	1,01	1,13	1,13	0.86	1,22	1,32

United States

Because of the low fuel taxes, the retail price of gasoline in the US is subject to greater fluctuations (vs. outside the US) when calculated as a percentage of cost-per-unit, but is less variable in absolute terms. From 1998 to 2004, the price of gasoline was between $1 and $2 USD per U.S. gallon. After 2004, the price increased until the average gas price reached a high of $4.11 per U.S. gallon in mid-2008, but has receded to approximately $2.60 per U.S. gallon as of September 2009. Recently, the U.S. has experienced an upswing in gas prices of 13.51% from Jan 31st to March 7, 2011.

Unlike most consumer goods, the prices of which are listed before tax, in the United States, gasoline prices are posted with taxes included. Taxes are added by federal, state and local governments. As of 2009, the federal tax is 18.4¢ per gallon for gasoline and 24.4¢ per gallon for diesel (excluding red diesel). Among states, the highest gasoline tax rates, including the federal taxes as of 2005, are New York (62.9¢/gal), Hawaii (60.1¢/gal), and California (60¢/gal). However, many states' taxes are a percentage and thus vary in amount depending on the cost of the gasoline.

About 9% of all gasoline sold in the US in May 2009 was premium grade, according to the Energy Information Administration. *Consumer Reports* magazine says, "If your car can run on regular, run it on regular." The Associated Press said premium gas—which is a higher octane and costs several cents a gallon more than regular unleaded—should be used only if the manufacturer says it is "required".

Etymology and Terminology

"Gasoline" is cited (under the spelling "gasolene") from 1865 in the *Oxford English Dictionary*. The trademark *Gasoline* was never registered, and eventually became generic in North America and the Philippines. The word "petrol" has been used in English to refer to raw petroleum since the 16th century. However, it was first used to refer to the refined fuel in 1892, when it was registered as a trade name by British wholesaler Carless, Capel & Leonard at the suggestion of Frederick Richard Simms, as a contraction of 'St. Peter's Oil'. Carless's competitors used the term "motor spirit" until the 1930s. The *Oxford English Dictionary* suggests this usage may have been inspired by the French *pétrole.*

In many countries, gasoline has a colloquial name derived from that of the chemical benzene (*e.g.*, German *Benzin*, Dutch *Benzine*). In other countries, especially in those portions of Latin America where Spanish predominates (*i.e.*, most of the region except Brazil), it has a colloquial name derived from that of the chemical naphtha (*e.g.*, Argentine/Uruguaian/Paraguaian *nafta*). However, the standard Spanish word is "gasolina."

The terms "mogas", short for motor gasoline, or "autogas", short for automobile gasoline, are used to distinguish automobile fuel from aviation gasoline, or "avgas". In British English, *gasoline* can refer to a different petroleum derivative historically used in lamps, but this usage is relatively uncommon.

Giant Oil and Gas Fields

The world's 932 giant oil and gas fields are considered those with 500 million barrels (79,000,000 m^3) of ultimately recoverable oil or gas equivalent. Geoscientists believe these giants account for 40 percent of the world's petroleum reserves. They are clustered in 27 regions of the world, with the largest clusters in the Persian Gulf and Western Siberian Basin. The past three decades reflect declines in discoveries of giant fields. The present decade (2000–2010), however, reflects an upturn in discoveries and appears on track to be the third best for discovery of giant oil and gas fields in the 150 year history of modern oil and gas exploration.

According to analysis led by Paul Mann of the University of Texas' Jackson School of Geosciences, almost all of the 932 giant oil and gas fields cluster within 27 regions, or about 30 percent of Earth's land surface. Since 2003, Mann and colleagues M.K. Horn and Ian Cross have tracked the giants on a map that highlights the tectonic and sedimentary basin maps of the 27 key regions. The map is in the public domain and available as a high-resolution pdf on the Web site of the Jackson School of Geosciences.

Recent work in tracking giant oil and gas fields follows the earlier efforts of the late exploration geologist Michel T. Halbouty, who tracked trends in giant discoveries from the 1960s to 2004.

Tectonic Settings

Geophysicists and exploration geologists who look for oil and gas fields classify the subsurface characteristics, or tectonic setting, of geological structures that contain hydrocarbons. Any one oil and gas field may reflect influences from multiple geological periods and events, but geoscientists often attempt to characterize a field based on the dominant geological event that influenced the structure's ability to trap and contain oil and gas in recoverable quantities.

A majority of the world's giant oil and gas fields exist in two characteristic tectonic settings—passive margin and rift environments. Passive margins are found along the edges of major ocean basins, such as the Atlantic coast of Brazil where oil and gas has been located in large quantities in the Campos basin. Rifts are oceanic ridges formed when tectonic plates separate and a new crust is created. The North Sea is an example of a rift setting associated with prodigious hydrocarbon reserves. Geoscientists theorize that both zones are especially conducive to forming giant oil and gas fields when they are distant from active tectonic areas. Stability appears to be conducive to trapping and retaining hydrocarbons under the subsurface.

Four other common tectonic settings, including collisional margins, strike-slip margins, and subduction margins, are associated with the formation of giant oil and gas fields, though not to the dominant extent of passive margin and rift settings.

Recent and Future Giants

Based on the locations of past giants, Mann et al. predicted new discoveries of giant oil and gas fields would mainly be made in passive margin and rift environments, especially in deepwater basins. They also predicted that existing areas that have produced giant fields would be likely targets for new discoveries of "elephants," as the fields are sometimes known in the oil and gas industry. Data from 2000-2007 reflect the accuracy of their predictions. The 79 new giant oil and gas fields discovered from 2000-2007 tended to be located in similar tectonic settings as the previously documented giants from 1868–2000, with 36 percent along passive margins, 30 percent in rift zones or overlying sags (structures associated with rifts), and 20 percent in collisional zones.

Despite a recent uptick in the number of giant oil and gas fields, discovery of giants appears to have peaked in the 1960s and 1970s. Looking to the future, geoscientists foresee a continuation of the recent trend of discovering more giant gas fields than oil fields. Two major continental regions—Antarctica and the Arctic—remain largely unexplored. Beyond them, however, trends suggest that remaining giant fields will be discovered in "in-fill" areas where past giants have been clustered and in frontier, or new, areas that correspond to the predominant tectonic settings of past giants.

Giant Field Production Properties and Behaviour

Comprehensive analysis of the production from the majority of the world's giant oil fields has shown their enormous importance for global oil production. For instance, the 20 largest oil fields in the world alone account for roughly 25% of the total oil production. The majority of the world's giants are already in decline and this will have dramatic consequences for future oil supply. Further analysis shows that giant oil fields typically reach their maximum production before 50% of the ultimate recoverable volume has been extracted. A strong correlation between depletion and the rate of decline was also found in that study, indicating that much new technology has only been able to temporarily decrease depletion at the expense of rapid future decline. This is exactly the case in the Cantarell Field.

Hydrocarbon Exploration

Hydrocarbon exploration (or oil and gas exploration) is the search by petroleum geologists and geophysicists for hydrocarbon deposits beneath the Earth's surface, such as oil and natural gas. Oil and gas exploration are grouped under the science of petroleum geology.

Exploration Methods

Visible surface features such as oil seeps, natural gas seeps, pockmarks (underwater craters caused by escaping gas) provide basic evidence of hydrocarbon generation (be it shallow or deep in the Earth). However, most exploration depends on highly sophisticated technology to detect and determine the extent of these deposits using exploration geophysics. Areas thought to contain hydrocarbons are initially subjected to a gravity survey, magnetic survey, passive seismic or regional seismic reflection surveys to detect large scale features of the sub-surface geology.

Features of interest (known as *leads*) are subjected to more detailed seismic surveys which work on the principle of the time it takes for reflected sound waves to travel through matter (rock) of varying densities and using the process of depth conversion to create a profile of the substructure. Finally, when a prospect has been identified and evaluated and passes the oil company's selection criteria, an exploration well is drilled in an attempt to conclusively determine the presence or absence of oil or gas.

Oil exploration is an expensive, high-risk operation. Offshore and remote area exploration is generally only undertaken by very large corporations or national governments. Typical shallow shelf oil wells (e.g. North Sea) cost USD$10 – 30 million, while deep water wells can cost up to USD$100 million plus. Hundreds of smaller companies search for onshore hydrocarbon deposits worldwide, with some wells costing as little as USD$100,000.

Elements of a Petroleum Prospect

A prospect is a potential trap which geologists believe may contain hydrocarbons. A significant amount of geological, structural and seismic investigation must first be completed to redefine the potential hydrocarbon drill location from a lead to a prospect. Five geological factors have to be present for a prospect to work and if any of them fail neither oil nor gas will be present.

- A source rock - When organic-rich rock such as oil shale or coal is subjected to high pressure and temperature over an extended period of time, hydrocarbons form.
- Migration - The hydrocarbons are expelled from source rock by three density-related mechanisms: the newly-matured hydrocarbons are less dense than their precursors, which causes overpressure; the hydrocarbons are lighter medium, and so migrate upwards due to buoyancy, and the fluids expand as further burial causes increased heating. Most hydrocarbons migrate to the surface as oil seeps, but some will get trapped.

- Trap - The hydrocarbons are buoyant and have to be trapped within a structural (e.g. Anticline, fault block) or stratigraphic trap
- Seal or cap rock - The hydrocarbon trap has to be covered by an impermeable rock known as a seal or cap-rock in order to prevent hydrocarbons escaping to the surface
- Reservoir - The hydrocarbons are contained in a reservoir rock. This is a porous sandstone or limestone. The oil collects in the pores within the rock. The reservoir must also be permeable so that the hydrocarbons will flow to surface during production.

Exploration Risk

Hydrocarbon exploration is a high risk investment and risk assessment is paramount for successful exploration portfolio management. Exploration risk is a difficult concept and is usually defined by assigning confidence to the presence of five imperative geological factors, as discussed above. This confidence is based on data and/or models and is usually mapped on Common Risk Segment Maps (CRS Maps). High confidence in the presence of imperative geological factors is usually colored green and low confidence colored red . Therefore these maps are also called Traffic Light Maps, while the full procedure is often referred to as Play Fairway Analysis . The aim of such procedures is to force the geologist to objectively assess all different geological factors. Furthermore it results in simple maps that can be understood by non-geologists and managers to base exploration decisions on.

Terms Used in Petroleum Evaluation

- Bright spot - On a seismic section, coda that have high amplitudes due to a formation containing hydrocarbons.
- Chance of success - An estimate of the chance of all the elements within a prospect working, described as a probability. High risk prospects have a less than 10% chance of working, medium risk prospects 10-20%, low risk prospects over 20%. Typically about 40% of wells recently drilled find commercial hydrocarbons.
- Dry hole - A formation that contains brine instead of oil.
- Flat spot - An oil-water contact on a seismic section; flat due to gravity.

- Hydrocarbon in place - amount of hydrocarbon likely to be contained in the prospect. This is calculated using the volumetric equation - GRV x N/G x Porosity x Sh x FVF
  - GRV - Gross rock volume - amount of rock in the trap above the hydrocarbon water contact
  - N/G - net/gross ratio - proportion of the GRV formed by the reservoir rock (range is 0 to 1)
  - Porosity - percentage of the net reservoir rock occupied by pores (typically 5-35%)
  - Sh - hydrocarbon saturation - some of the pore space is filled with water - this must be discounted
  - FVF - formation volume factor - oil shrinks and gas expands when brought to the surface. The FVF converts volumes at reservoir conditions (high pressure and high temperature) to storage and sale conditions
- Lead - a structure which may contain hydrocarbons
- Play - A particular combination of reservoir, seal, source and trap associated with proven hydrocarbon accumulations
- Prospect - a lead which has been fully evaluated and is ready to drill
- Recoverable hydrocarbons - amount of hydrocarbon likely to be recovered during production. This is typically 10-50% in an oil field and 50-80% in a gas field.

Licensing

Petroleum resources are typically owned by the government of the host country. In the USA most onshore (land) oil and gas rights (OGM) are owned by private individuals. Sometimes this is not the same person who owns the surface rights. In this case oil companies must negotiate terms for a lease of these rights with the individual who owns the OGM. In most nations the government issues licences to explore, develop and produce its oil and gas resources, which are typically administered by the oil ministry. There are several different types of licence. Typically oil companies operate in joint ventures to spread the risk, one of the companies in the partnership is designated the operator who actually supervises the work.

- Tax and Royalty - Companies would pay a royalty on any oil produced, together with a profits tax (which can have expenditure offset against it). In some cases there are also various bonuses and ground rents (license fees) payable to the government - for example a signature bonus payable at the start of the licence. Licences are awarded in competitive bid rounds on the basis of either the size of the work programme (number of wells, seismic etc.) or size of the signature bonus.

- Production Sharing contract (PSA) - A PSA is more complex than a Tax/Royalty system - The companies bid on the percentage of the production that the host government receives (this may be variable with the oil price), There is often also participation by the Government owned National Oil Company (NOC). There are also various bonuses to be paid. Development expenditure is offset against production revenue.
- Service contract - This is when an oil company acts as a contractor for the host government, being paid to produce the hydrocarbons.

Reserves and Resources

Resources are hydrocarbons which may or may not be produced in the future. A resource number may be assigned to an undrilled prospect or an unappraised discovery.

Appraisal by drilling additional delineation wells or acquiring extra seismic data will confirm the size of the field and lead to project sanction. At this point the relevant government body gives the oil company a production licence which enables the field to be developed. This is also the point at which oil reserves can be formally booked.

Oil reserves are primarily a measure of geological risk - of the probability of oil existing and being producible under current economic conditions using current technology.

The three categories of reserves generally used are proven, probable, and possible reserves.

- Proven reserves - defined as oil and gas "Reasonably Certain" to be producible using current technology at current prices, with current commercial terms and government consent- also known in the industry as 1P. Some Industry specialists refer to this as P90 - i.e. having a 90% certainty of being produced.
- Probable reserves - defined as oil and gas "Reasonably Probable" of being produced using current or likely technology at current prices, with current commercial terms and government consent

- Some Industry specialists refer to this as P50 - i.e. having a 50% certainty of being produced. - This is also known in the industry as 2P or Proven plus probable.

- Possible reserves - i.e. "having a chance of being developed under favourable circumstances" - Some industry specialists refer to this as P10 - i.e. having a 10% certainty of being produced. - This is also known in the industry as 3P or Proven plus probable plus possible.

Reserve Booking

Oil and gas reserves are the main asset of an oil company - booking is the process by which they are added to the Balance sheet. This is done according to a set of rules developed by the Society of Petroleum Engineers (SPE). The Reserves of any company listed on the New York Stock Exchange have to be stated to the U.S. Securities and Exchange Commission. In many cases these reported reserves are audited by external geologists, although this is not a legal requirement. The U.S. Securities and Exchange Commission rejects the probability concept and prohibits companies from mentioning probable and possible reserves in their filings. Thus, official estimates of proven reserves will always be understated compared to what oil companies think actually exists. For practical purposes companies will use proven plus probable estimate (2P), and for long term planning they will be looking primarily at possible reserves.

Other countries also have their national hydrocarbon reserves authorities for example, Russia's State Commission on Mineral Reserves (GKZ), to which companies operating in these countries have to report.

Countries by Proven Oil Reserves

Proved reserves are those quantities of petroleum which, by analysis of geological and engineering data, can be estimated with a high degree of confidence to be commercially recoverable from a given date forward, from known reservoirs and under current economic conditions.Several Oil Producers which are not considered to be countries (Eg the European Union, and The Arab league) are included in the list because they provide useful reference information and are included in some of the sources.

These are not ranked in the charts here,

	Country	*Reserves (bbl)*
	Arab League (more information) (2011)	688,860,600,000
1	Venezuela (more information) (2010)	296,500,000,000
2	Saudi Arabia (more information) (2011)	264,600,000,000
3	Canada (more information) (2008)	175,200,000,000
4	Iran (more information) (2006)	137,600,000,000
5	Iraq (more information) (2008)	115,000,000,000
6	Kuwait (more information) (2010)	104,000,000,000
7	United Arab Emirates (more information) (2008)	97,800,000,000
8	Russia (more information) (2009)	74,200,000,000
9	Libya (more information) (2010)	47,000,000,000
10	Nigeria (more information) (2007)	37,500,000,000
11	Kazakhstan (2009)	30,000,000,000
12	Qatar (2009)	25,410,000,000
13	China	20,350,000,000
14	United States (more information)	19,120,000,000
15	Azerbaijan	14,000,000,000
16	Brazil	13,986,000,000
17	Angola	13,500,000,000
18	Mexico (more information)	12,420,000,000
19	Algeria	12,200,000,000
20	Sudan	6,800,000,000
21	Norway	6,680,000,000
22	Ecuador	6,542,000,000
23	India	5,800,000,000
24	Oman	5,500,000,000
—	European Union	5,453,000,000
25	Ghana	5,000,000,000
26	Vietnam	4,700,000,000
27	Egypt	4,300,000,000
28	Indonesia	4,050,000,000
29	Gabon	3,700,000,000
30	Australia	1,100,000,000
31	United Kingdom	3,000,000,000
32	Yemen	3,000,000,000

Contd...

	Country	***Reserves (bbl)***
33	Malaysia	2,900,000,000
34	Syria	2,500,000,000
35	Argentina	2,386,000,000
36	Colombia	1,900,000,000
37	Congo, Republic of the	1,600,000,000
38	Chad	1,500,000,000
39	Brunei	1,100,000,000
40	Equatorial Guinea	1,100,000,000
41	Denmark	1,060,000,000
42	Trinidad and Tobago	728,300,000
43	Romania	600,000,000
44	Turkmenistan	600,000,000
45	Uzbekistan	594,000,000
46	Timor-Leste	553,800,000
47	Peru	470,800,000
48	Bolivia	465,000,000
49	Pakistan	436,200,000
50	Thailand	430,000,000
51	Tunisia	425,000,000
52	Italy	423,700,000
53	Ukraine	395,000,000
54	Germany	276,000,000
55	Turkey	262,200,000
56	Cote d'Ivoire	250,000,000
57	Cameroon	200,000,000
58	Albania	199,100,000
59	Belarus	198,000,000
60	Congo, Democratic Republic of the	180,000,000
61	Cuba (more information)	178,900,000
62	Papua New Guinea	170,000,000
63	Philippines	168,000,000
64	Chile	150,000,000
65	Spain	150,000,000

Contd...

	Country	**Reserves (bbl)**
66	Bahrain	124,600,000
67	France	101,200,000
68	Mauritania	100,000,000
69	Netherlands	100,000,000
70	Morocco	100,000,000
71	Poland	96,380,000
72	Austria	89,000,000
73	Guatemala	83,070,000
74	Suriname	79,600,000
75	Serbia	77,500,000
76	Croatia	73,350,000
77	New Zealand	60,000,000
78	Myanmar	50,000,000
79	Japan	44,120,000
80	Kyrgyzstan	40,000,000
81	Georgia	35,000,000
82	Bangladesh	28,000,000
83	Hungary	26,570,000
84	Bulgaria	15,000,000
85	South Africa	15,000,000
86	Czech Republic	15,000,000
87	Lithuania	12,000,000
88	Tajikistan	12,000,000
89	Greece	10,000,000
90	Slovakia	9,000,000
91	Benin	8,000,000
92	Belize	6,700,000
93	Taiwan	2,800,000
94	Israel	1,940,000
95	Barbados	1,790,000
96	Jordan	1,000,000
97	Ethiopia	430,000
-	Total	1,392,461,050,000

6

Oil Shale Reserves

Oil shale reserves refers to oil shale resources that are recoverable under given economic restraints and technological abilities. Oil shale deposits range from small presently non-economic occurrences to large presently commercially exploitable reserves.

Defining oil shale reserves is difficult, as the chemical composition of different oil shales, as well as their kerogen content and extraction technologies, vary significantly.

The economic feasibility of shale oil extraction is highly dependent on the price of conventional oil; if the price of crude oil per barrel is less than the production price per barrel of shale oil, it is uneconomic.

There are around 600 known oil shale deposits. Many deposits need more exploration to determine their potential as reserves.

However, worldwide technically recoverable reserves have recently been estimated at about 2.8–3.3 trillion barrels ($450\times10^{\triangle 9}$–$520\times10^{\triangle 9}$ m^3) of shale oil, with the largest reserves in the United States, which is thought to have 1.5–2.6 trillion barrels ($240\times10^{\triangle 9}$–$410\times10^{\triangle 9}$ m^3).

Well-explored deposits, which could be classified as reserves, include the Green River deposits in the western United States, the Tertiary deposits in Queensland, Australia, deposits in Sweden and Estonia, the El-Lajjun deposit in Jordan, and deposits in France, Germany, Brazil, China, and Russia.

It is expected that these deposits would yield at least 40 liters (0.25 bbl) of shale oil per metric ton of shale, using the Fischer Assay.

Definition of Reserves

Estimating shale oil reserves is complicated by several factors. Firstly, the amount of kerogen contained in oil shale deposits varies considerably.

Secondly, some nations report as reserves the total amount of kerogen in place, including all kerogen regardless of technical or economic constraints; these estimates do not consider the amount of kerogen that may be extracted from identified and assayed oil shale rock using available technology and under given economic conditions. By most definitions, "reserves" refers only to the amount of resource which is technically exploitable and economically feasible under current economic conditions.

The term "resources", on the other hand, may refer to all deposits containing kerogen. Thirdly, shale oil extraction technologies are still developing, so the amount of recoverable kerogen can only be estimated.

There are a wide variety of extraction methods, which yield significantly different quantities of useful oil. As a result, the estimated amounts of resources and reserves display wide variance.

The kerogen content of oil shale formations differs widely, and the economic feasibility of its extraction is highly dependent on international and local costs of oil.

Several methods are used to determine the quantity and quality of the products extracted from shale oil. At their best, these methods give an approximate value to its energy potential.

One standard method is the Fischer Assay, which yields a heating value, that is, a measure of caloric output. This is generally considered a good overall measure of usefulness.

The Fischer Assay has been modified, standardized, and adapted by the American Petroleum Institute. It does not, however, indicate how much oil could be extracted from the sample.

Some processing methods yield considerably more useful product than the Fischer Assay would indicate.

The Tosco II method yields over 100% more oil, and the Hytort process yields between 300% to 400% more oil.

Table: *Largest oil shale deposits (over 1 billion metric tons*

Deposit	***Country***	***Period***	***In-place shale oil resources (million barrels)***	***In-place oil shale resources (million metric tons)***
Green River Formation	USA	Tertiary	1,466,000	213,000
Phosphoria Formation	USA	Permian	250,000	35,775
Eastern Devonian	USA	Devonian	189,000	27,000
Heath Formation	USA	Early Carboniferous	180,000	25,578
Olenyok Basin	Russia	Cambrian	167,715	24,000
Congo	Democratic Republic of Congo	?	100,000	14,310
Irati Formation	Brazil	Permian	80,000	11,448
Sicily	Italy	?	63,000	9,015
Tarfaya	Morocco	Cretaceous	42,145	6,448
Volga Basin	Russia	?	31,447	4,500
St. Petersburg, Baltic Oil Shale Basin	Russia	Ordovician	25,157	3,600
Vychegodsk Basin	Russia	Jurassic	19,580	2,800
Wadi Maghar	Jordan	Cretaceous	14,009	2,149
Dictyonema shale	Estonia	Ordovician	12,386	1,900
Timahdit	Morocco	Cretaceous	11,236	1,719
Collingwood Shale	Canada	Ordovician	12,300	1,717
Italy	Italy	Triassic	10,000	1,431

Geographical Allocation

There is no comprehensive overview of oil shales geographical allocation around the world. Around 600 known oil shale deposits are diversely spread throughout the earth, and are found on every continent with the possible exception of Antarctica, which has not yet been explored for oil shale.

Oil shale resources can be concentrated in a large confined deposit such as the Green River formations, which were formed by a large inland lake. These can be many meters thick but limited by the size of the original lake. They may also resemble the deposits found along the eastern American seaboard, which were the product of a shallow sea, in that they may be quite thin but laterally expansive, covering thousands of square kilometers.

The table below reports reserves by estimated amount of shale oil. Shale oil refers to synthetic oil obtained by heating organic material (kerogen) contained in oil shale to a temperature which will separate it into oil, combustible gas, and the residual carbon that remains in the spent shale. All figures are presented in barrels and metric tons.

Shale oil: resources and production at end-2005 by regions and countries with resources over 10 billion barrels (1.6 km^3) of in-place shale oil.

Region	***In-place shale oil resources (million barrels)***	***In-place oil shale oil resources (million barrels)***	***Production in 2002 (thousand metric tons (oil))***
Africa	159,243	23,317	-
Democratic Republic of the Congo	100,000	14,310	-
Morocco	53,381	8,187	-
Asia	45,894	6,562	180
China	16,000	2,290	180
Europe	368,156	52,845	345
Russia	247,883	35,470	-
Italy	73,000	10,446	-
Estonia	16,286	2,494	345
Middle East	38,172	5,792	-
Jordan	34,172	5,242	-
North America	2,100,469	303,758	-
United States	2,085,228	301,566	-
Canada	15,241	2,192	-
Oceania	31,748	4,534	-
Australia	31,729	4,531	-
South America	82,421	11,794	157
Brazil	82,000	11,734	159
World total	2,826,103	408,602	684

Africa

Major oil shale deposits are located in the Democratic Republic of Congo (equal to 14.31 billion metric tons of shale oil) and Morocco (12.3 billion metric tons or 8.16 billion metric tons of shale oil). Deposits in Congo are not properly explored yet. In Morocco, oil shale deposits have been identified at ten localities with the largest deposits in Tarfaya and Timahdite.

Although reserves in Tarfaya and Timahdit are well explored, the commercial exploitation has not started yet and only a limited program of laboratory and pilot-plant research has been undertaken. There are also oil shale reserves in Egypt, South Africa, Madagascar, and Nigeria. The main deposits of Egypt are located in Safaga-Al-Qusair and Abu Tartour areas.

Asia

Major oil shale deposits are located in China, which has an estimated total of 32 billion metric tons, of which 4.4 billion metric tons are technically exploitable and economically feasible; Thailand (18.7 billion metric tons), Kazakhstan (several deposits; major deposit at Kenderlyk Field with 4 billion metric tons), and Turkey (2.2 billion metric tons). Smaller oil shale reserves have also been found in Assam (India), Pakistan, Uzbekistan, Turkmenistan, Myanmar, Armenia, and Mongolia.

The principal Chinese oil shale deposits and production lie in Fushun and Liaoning; others are located in Maoming in Guangdong, Huadian in Jilin, Heilongjiang, and Shandong. In 2002, China produced more than 90,000 metric tons of shale oil. Thailand's oil shale deposits are near Mae Sot, Tak Province, and at Li, Lamphun Province. Deposits in Turkey are found mainly in middle and western Anatolia.

Professor Alan R. Carroll of University of Wisconsin–Madison estimates that Upper Permian lacustrine oil shale deposits of northwest China, absent from previous global oil shale assessments, are comparable to the Green River Formation.

Europe

The biggest oil shale reserves in Europe are located in Russia (equal to 35.47 billion metric tons of shale oil). Major deposits are located in the Volga-Petchyorsk province and in the Baltic Oil Shale Basin.

Figure: *Outcrop of Ordovician kukersite oil shale, northern Estonia.*

Other major oil shale deposits in Europe are located in Italy (10.45 billion metric tons of shale oil), Estonia (2.49 billion metric tons of shale oil), France (1 billion metric tons of shale oil), Belarus (1 billion metric tons of shale oil), Sweden (875 million metric tons of shale oil), Ukraine (600 million metric tons of shale oil) and the United Kingdom (500 million metric tons of shale oil). There are oil shale reserves also in Germany, Luxembourg, Spain, Bulgaria, Hungary, Poland, Serbia, Austria, Albania, and Romania.

Middle East

Significant oil shale deposits are located in Jordan (5.242 billion metric tons of shale oil or 65 billion metric tons of oil shale) and Israel (550 million metric tons of shale oil or 6.5 billion metric tons of oil shale). Jordanian oil shales are high quality, comparable to western US oil shale, although their sulfur content is high. The best-explored deposits are El Lajjun, Sultani, and the Juref ed Darawishare located in west-central Jordan, while the Yarmouk deposit, close to its northern border, extends into Syria. Most of Israel's deposits are located in the Rotem Basin region of the northern Negev desert near the Dead Sea. Israeli oil shale is relatively low in heating value and oil yield.

North America

At 301 billion metric tons, the oil shale deposits in the United States are easily the largest in the world. There are two major deposits:

the eastern US deposits, in Devonian-Mississippian shales, cover 250,000 square miles (650,000 km^2); the western US deposits of the Green River Formation in Colorado, Wyoming, and Utah, are among the richest oil shale deposits in the world. In Canada 19 deposits have been identified. The best-examined deposits are in Nova Scotia and New Brunswick.

Oceania

Australia's oil shale resource is estimated at about 58 billion metric tons or 4.531 billion metric tons of shale oil, of which about 24 billion barrels (3.8 billion cubic metres) is recoverable. The deposits are located in the eastern and southern states with the biggest potential in the eastern Queensland deposits. Oil shale has also been found in New Zealand.

South America

Brazil has the world's second-largest known oil shale resources (the Irati shale and lacustrine deposits) and is currently the world's second largest shale oil producer, after Estonia. Oil shale resources occur in São Mateus do Sul, Paraná, and in Vale do Paraiba. Brazil has developed the world's largest surface oil shale pyrolysis retort at Petrosix, with a 11-meter (36 ft)-diameter vertical shaft. Brazilian production in 1999 was about 200,000 metric tons. Small resources are also found in Argentina, Chile, Paraguay, Peru, Uruguay, and Venezuela.

OPEC

OPEC (Organization of Petroleum Exporting Countries) is an intergovernmental organization of twelve developing countries made up of Algeria, Angola, Ecuador, Iran, Iraq, Kuwait, Libya, Nigeria, Qatar, Saudi Arabia, the United Arab Emirates, and Venezuela. OPEC has maintained its headquarters in Vienna since 1965, and hosts regular meetings among the oil ministers of its Member Countries. Indonesia withdrew in 2008 after it became a net importer of oil, but stated it would likely return if it became a net exporter again.

According to its statutes, one of the principal goals is the determination of the best means for safeguarding the organization's interests, individually and collectively. It also pursues ways and means of ensuring the stabilization of prices in international oil markets with a view to eliminating harmful and unnecessary fluctuations; giving due regard at all times to the interests of the producing nations and to the necessity of securing a steady income to the producing countries; an efficient and regular supply of petroleum to consuming nations, and a fair return on their capital to those investing in the petroleum industry.

OPEC's influence on the market has been widely criticized, since it became effective in determining production and prices. Arab members of OPEC alarmed the developed world when they used the "oil weapon" during the Yom Kippur War by implementing oil embargoes and initiating the 1973 oil crisis. Although largely political explanations for the timing and extent of the OPEC price increases are also valid, from OPEC's point of view, these changes were triggered largely by previous unilateral changes in the world financial system and the ensuing period of high inflation in both the developed and developing world. This explanation encompasses OPEC actions both before and after the outbreak of hostilities in October 1973, and concludes that "OPEC countries were only 'staying even' by dramatically raising the dollar price of oil."

OPEC's ability to control the price of oil has diminished somewhat since then, due to the subsequent discovery and development of large oil reserves in Alaska, the North Sea, Canada, the Gulf of Mexico, the opening up of Russia, and market modernization. As of November 2010, OPEC members collectively hold 79% of world crude oil reserves and 44% of the world's crude oil production, affording them considerable control over the global market. The next largest group of producers, members of the OECD and the Post-Soviet states produced only 23.8% and 14.8%, respectively, of the world's total oil production. As early as 2003, concerns that OPEC members had little excess pumping capacity sparked speculation that their influence on crude oil prices would begin to slip.

History

Venezuela and Iran were the first countries to move towards the establishment of OPEC in the 1960s by approaching Iraq, Kuwait and Saudi Arabia in 1949, suggesting that they exchange views and explore avenues for regular and closer communication among petroleum-producing nations. The founding members are Iran, Iraq, Kuwait, Saudi Arabia, and Venezuela. Later members include Algeria, Ecuador, Gabon, Indonesia, Libya, Qatar, Nigeria, and the United Arab Emirates.

In 10–14 September 1960, at the initiative of the Venezuelan Energy and Mines minister Juan Pablo Pérez Alfonzo and the Saudi Arabian Energy and Mines minister Abdullah al-Tariki, the governments of Iraq, Iran, Kuwait, Saudi Arabia and Venezuela met in Baghdad to discuss ways to increase the price of the crude oil produced by their respective countries.

OPEC was founded to unify and coordinate members' petroleum policies. Original OPEC members include Iran, Iraq, Kuwait, Saudi Arabia, and Venezuela. Between 1960 and 1975, the organization expanded to include Qatar (1961), Indonesia (1962), Libya (1962), the United Arab Emirates (1967), Algeria (1969), and Nigeria (1971). Ecuador and Gabon were early members of OPEC, but Ecuador withdrew on December 31, 1992 because it was unwilling or unable to pay a $2 million membership fee and felt that it needed to produce more oil than it was allowed to under the OPEC quota, although it rejoined in October 2007. Similar concerns prompted Gabon to suspend membership in January 1995. Angola joined on the first day of 2007. Norway and Russia have attended OPEC meetings as observers. Indicating that OPEC is not averse to further expansion, Mohammed Barkindo, OPEC's Secretary General, recently asked Sudan to join. Iraq remains a member of OPEC, but Iraqi production has not been a part of any OPEC quota agreements since March 1998.

In May 2008, Indonesia announced that it would leave OPEC when its membership expired at the end of that year, having become a net importer of oil and being unable to meet its production quota. A statement released by OPEC on 10 September 2008 confirmed Indonesia's withdrawal, noting that it "regretfully accepted the wish of Indonesia to suspend its full Membership in the Organization and recorded its hope that the Country would be in a position to rejoin the Organization in the not too distant future." Indonesia is still exporting light, sweet crude oil and importing heavier, more sour crude oil to take advantage of price differentials (import is greater than export) due to Air pollution in Indonesia still being low as compared to China or The United States.

1973 Oil Embargo

The persistence of the Arab-Israeli conflict finally triggered a response that transformed OPEC into a formidable political force. After the Six Day War of 1967, the Arab members of OPEC formed a separate, overlapping group, the Organization of Arab Petroleum Exporting Countries, for the purpose of centering policy and exerting pressure on the West over its support of Israel. Egypt and Syria, though not major oil-exporting countries, joined the latter grouping to help articulate its objectives. Later, the Yom Kippur War of 1973 galvanized Arab opinion. Furious at the emergency re-supply effort that had enabled Israel to withstand Egyptian and Syrian forces, the Arab world imposed the 1973 oil embargo against the United States and Western Europe, while non-Arab OPEC members did not..

1975 Hostage Incident

On 21 December 1975 Ahmed Zaki Yamani and the other oil ministers of the members of OPEC were taken hostage by a six-person team led by terrorist Carlos the Jackal (which included Gabriele Krocher-Tiedemann and Hans-Joachim Klein), in Vienna, Austria, where the ministers were attending a meeting at the OPEC headquarters. Carlos planned to take over the conference by force and kidnap all eleven oil ministers in attendance and hold them for ransom, with the exception of Ahmed Zaki Yamani and Iran's Jamshid Amuzegar, who were to be executed.

Carlos led his six-person team past two police officers in the building's lobby and up to the first floor, where a police officer, an Iraqi plain clothes security guard and a young Libyan economist were shot dead.

As Carlos entered the conference room and fired shots in the ceiling, the delegates ducked under the table. The terrorists searched for Ahmed Zaki Yamani and then divided the sixty-three hostages into groups. Delegates of friendly countries were moved toward the door, 'neutrals' were placed in the centre of the room and the 'enemies' were placed along the back wall, next to a stack of explosives. This last group included those from Saudi Arabia, Iran, Qatar and the UAE. Carlos demanded a bus to be provided to take his group and the hostages to the airport, where a DC-9 airplane and crew would be waiting. In the meantime, Carlos briefed Ahmed Zaki Yamani on his plan to eventually fly to Aden, where Yamani and the Iranian minister would be killed.

The bus was provided the following morning at 6.40 as requested and 42 hostages were boarded and taken to the airport. The group was airborne just after 9.00 and explosives placed under Yamani's seat. The plane first stopped in Algiers, where Carlos left the plane to meet with the Algierian Foreign minister. All 30 non-Arab hostages were released, excluding Amuzegar.

The refueled plane left for Tripoli where there was trouble in acquiring another plane as had been planned. Carlos decided to instead return to Algiers and change to a Boeing 707, a plane large enough to fly to Baghdad nonstop. Ten more hostages were released before leaving.

With only 10 hostages remaining, the Boeing 707 left for Algiers and arrived at 3.40 a.m. After leaving the plane to meet with the Algerians, Carlos talked with his colleagues in the front cabin of the plane and then told Yamani and Amouzegar that they would be released at mid-day. Carlos was then called from the plane a second time and returned after two hours.

At this second meeting it is believed that Carlos held a phone conversation with Algerian President Houari Boumédienne who informed Carlos that the oil ministers' deaths would result in an attack on the plane. Yamani's biography suggests that the Algerians had used a covert listening device on the front of the aircraft to overhear the earlier conversation between the terrorists, and found that Carlos had in fact still planned to murder the two oil ministers. Boumédienne must also have offered Carlos asylum at this time and possibly financial compensation for failing to complete his assignment.

On returning to the plane Carlos stood before Yamani and Amuzegar and expressed his regret at not being able to murder them. He then told the hostages that he and his comrades would leave the plane after which they would all be free. After waiting for the terrorists to leave, Yamani and the other nine hostages followed and were taken to the airport by Algerian Foreign Minister Abdelaziz Bouteflika. The terrorists were present in the next lounge and Khalid, the Palestinian, asked to speak to Yamani. As his hand reached for his coat, Khalid was surrounded by guards and a gun was found concealed in a holster.

Some time after the attack it was revealed by Carlos' accomplices that the operation was commanded by Wadi Haddad, a Palestinian terrorist and founder of the Popular Front for the Liberation of Palestine. It was also claimed that the idea and funding came from an Arab president, widely thought to be Muammar al-Gaddafi.

In the years following the OPEC raid, Bassam Abu Sharif and Klein claimed that Carlos had received a large sum of money in exchange for the safe release of the Arab hostages and had kept it for his personal use.

There is still some uncertainty regarding the amount that changed hands but it is believed to be between US$20 million and US$50 million.

The source of the money is also uncertain, but, according to Klein, it was from "an Arab president." Carlos later told his lawyers that the money was paid by the Saudis on behalf of the Iranians and was, "diverted en route and lost by the Revolution".

The 1980s Oil Gluts

After 1980, oil prices began a six-year decline that culminated with a 46 percent price drop in 1986.

This was due to reduced demand and over-production that produced a glut on the world market. Around this period, Iraq also increased its oil production to help pay for the Iran-Iraq War. Overall OPEC lost its unity and thus its net oil export revenues fell in the 1980s.

Responding to War and Low Prices

Leading up to the 1990-91 Gulf War, Iraqi President Saddam Hussein advocated that OPEC push world oil prices up, thereby helping Iraq, and other member states, service debts.

But the division of OPEC countries occasioned by the Iraq-Iran War and the Iraqi invasion of Kuwait marked a low point in the cohesion of OPEC. Once supply disruption fears that accompanied these conflicts dissipated, oil prices began to slide dramatically.

After oil prices slumped at around $15 a barrel in the late 1990s, concerted diplomacy, sometimes attributed to Venezuela's president Hugo Chávez, achieved a coordinated scaling back of oil production beginning in 1998.

In 2000, Chávez hosted the first summit of heads of state of OPEC in 25 years. The next year, however, the September 11, 2001 attacks against the United States, the following invasion of Afghanistan, and 2003 invasion of Iraq and subsequent occupation prompted a surge in oil prices to levels far higher than those targeted by OPEC during the preceding period.

Indonesia withdrew from OPEC to protect its oil supply interests. On November 19, 2007, global oil prices reacted strongly as OPEC members spoke openly about potentially converting their cash reserves to the euro and away from the US dollar.

Production Disputes

The economic needs of the OPEC member states often affects the internal politics behind OPEC production quotas. Various members have pushed for reductions in production quotas to increase the price of oil and thus their own revenues.

These demands conflict with Saudi Arabia's stated long-term strategy of being a partner with the world's economic powers to ensure a steady flow of oil that would support economic expansion. Part of the basis for this policy is the Saudi concern that expensive oil or oil of uncertain supply will drive developed nations to conserve and develop alternative fuels. To this point, former Saudi Oil Minister Sheikh Yamani famously said in 1973: "The stone age didn't end because we ran out of stones."

One such production dispute occurred on September 10, 2008, when the Saudis reportedly walked out of OPEC negotiating session where the organization voted to reduce production. Although Saudi Arabian OPEC delegates officially endorsed the new quotas, they stated anonymously that they would not observe them. The New York Times quoted one such anonymous OPEC delegate as saying "Saudi Arabia will meet the market's demand. We will see what the market requires and we will not leave a customer without oil. The policy has not changed."

OPEC Aid

OPEC aid dates from well before the 1973/74 oil price explosion. Kuwait has operated a programme since 1961 (through the Kuwait Fund for Arab Economic Development). The OPEC fund became a fully fledged permanent international development agency in May 1980.

Membership

Current members

OPEC has twelve member countries: six in the Middle East, four in Africa, and two in South America.

Country	*Region*	*Joined OPEC*	*Population (July 2008)*	*Area (km²)*
Algeria	Africa	1969	33,779,668	2,381,740
Angola	Africa	2007	12,531,357	1,246,700
Ecuador	South America	2007	13,927,650	283,560
Iran	Middle East	1960	75,875,224	1,648,000
Iraq	Middle East	1960	28,221,180	437,072
Kuwait	Middle East	1960	2,596,799	17,820
Libya	Africa	1962	6,173,579	1,759,540
Nigeria	Africa	1971	158,259,000	923,768
Qatar	Middle East	1961	824,789	11,437
Saudi Arabia	Middle East	1960	28,146,656	2,149,690
United Arab Emirates	Middle East	1967	4,621,399	83,600
Venezuela	South America	1960	26,414,816	912,050
Total			369,368,429	11,854,977 km²

1. Ecuador initially joined in 1973, left in 1992, and rejoined in 2007.

2. One of five founder members that attended the first OPEC conference, in September 1960.

Former members

Country	*Region*	*Joined OPEC*	*Left OPEC*
Gabon	Africa	1975	1994
Indonesia	South East Asia	1962	2009

The United States was a de facto member during its formal occupation of Iraq via the Coalition Provisional Authority.

Indonesia left OPEC in 2008 because it ceased to be a net exporter of oil. It could not fulfill the demand of its own country's needs, as growth in demand outstripped output.

The situation was made worse because of weak legal certainty and corruption that deterred foreign investors from investing in new reserves in Indonesia.

In recent times, the government has increased financial incentives for foreign firms to invest in exploration and extraction but has found itself forced to import more supplies from the likes of Iran, Saudi Arabia and Kuwait. Indonesia's departure from OPEC will not likely affect the amount of oil it produces or imports. The country's growing dependence on imports is proving increasingly expensive as global prices soar.

Economics

OPEC is a swing producer and its decisions have had considerable influence on international oil prices. For example, in the 1973 energy crisis OPEC refused to ship oil to western countries that had supported Israel in the Yom Kippur War, which Israel had fought against Egypt and Syria.

This refusal caused a fourfold increase in the price of oil, which lasted five months, starting on October 17, 1973, and ending on March 18, 1974. OPEC nations then agreed, on January 7, 1975, to raise crude oil prices by 10%.

At that time, OPEC nations — including many who had recently nationalized their oil industries —joined the call for a new international economic order to be initiated by coalitions of primary producers.

Concluding the First OPEC Summit in Algiers they called for stable and just commodity prices, an international food and agriculture program, technology transfer from North to South, and the democratization of the economic system.

Overall, the evidence suggests that OPEC did act as a cartel when it adopted output rationing in order to maintain price.

According to US government, in 2011 OPEC will break above the $1 trillion mark earnings for the first time at $1.034 trillion and it is beating the $965 billion peak set in 2008.

Sustainability

According to Mikael Hook, who researches the life cycles of oil fields, despite technological advances that increase the productivity of oil wells, the rate of decline of oil fields will eventually increase as time continues.

Energy policy expert Joyce Dargay accuses OPEC, along with several other institutions, of drastically under predicting future oil demand by 2030 by more than 25%, a difference of 28 million barrels per day (4,500,000 m^3/d) or about twice the current amount supplied by Saudi Arabia.

Quotas circa 2005

Table 2: OPEC Quotas and Production in thousands of barrels per day

Country	*Quota (7/1/05)*	*Production (1/07)*	*Capacity*
Saudi Arabia	10,099	9,800	12,500
Algeria	894	1,360	1,430
Angola	1,900	1,700	1,700
Ecuador	520	500	500
Iran	4,110	3,700	3,750
Iraq		1,481	
Kuwait	2,247	2,500	2,600
Libya	1,500	1,650	1,700
Nigeria	2,306	2,250	2,250
Qatar	726	810	850
United Arab Emirates	2,444	2,500	2,600
Venezuela	3,225	2,340	2,450
Total	29,971	29,591	30,330

Organization of Arab Petroleum Exporting Countries

The Organization of Arab Petroleum Exporting Countries (OAPEC) is a multi-governmental organization headquartered in Kuwait which coordinates energy policies between oil–producing Arab nations, and whose main purpose is developmental.

History

On 9 January 1968, three of the then–most conservative Arab oil states Kuwait, Libya, and Saudi Arabia agreed at a conference in Beirut, Lebanon to found the Organization of Arab Petroleum Engineering Countries, aiming to separate the production and sale of oil from politics in the wake of the halfhearted 1967 oil embargo in response to the Six Day War. Such use of the economic weapon of oil embargo in the struggle against Israel had been regularly proposed at Arab Petroleum Congresses, but it took the Six Day War for the embargo happen. However Saudi Arabia's oil production was up by 9% that year, and the main embargo lasted only ten days and was completely ended by the Khartoum Conference.

OAPEC was originally intended to be a conservative Arab political organization which, by restricting membership to countries whose main export was oil, would exclude governments seen as radical such as Egypt and Algeria. This organizational exclusivity was bolstered by an additional rule in the organization's charter requiring the three founders' approval all new members.

The original aim was to control the economic weapon of potential oil embargo and prevent its use caused by popular emotion. Iraq initially declined to join, preferring to work under the umbrella of the Arab League, considering OAPEC too conservative.

Equally the three founders considered Iraq too radical to be desirable as a member. However, by early 1972, the criteria for admission changed to oil being a significant source, rather than the principal source of revenue of a prospective member nation and Algeria, Iraq, Syria and Egypt had been admitted. Consequently the OPAEC became a much more activist organization, contrary to the original intention.

1973 was a turning point for the organization. In October that year, the forces of Egypt and Syria attempted to overwhelm the state of Israel in an offensive later known as the Yom Kippur War. On October 16, ten days after the war's start, Kuwait hosted separate meetings of both OAPEC and the Persian Gulf members of OPEC, including Iran. OAPEC resolved to cut oil production 5% monthly "until the Israeli forces are completely evacuated from all the Arab territories occupied in the June 1967 war...." The embargo would last for some five months before it was lifted in March 1974 after negotiations at the Washington Oil Summit.

The embargo's aftereffects would linger throughout the rest of the decade. For the oil exporting countries, the embargo was the first instance of the exercise of their ability to leverage their production for political gains. A number of the member nations would use this sense of control to renegotiate the contracts they had made with the companies that had discovered and exploited their resources. Ironically the vastly increased revenues would prove addictive, and a unified OAPEC oil embargo was never again possible.

In 1979, Egypt was expelled from OAPEC for signing the Camp David Accords, although it was readmitted a decade later.

OAPEC is regarded as a regional, specialized international organization focusing on organizing cooperation in oil development, collective projects, and regional integration.

Members

- Saudi Arabia (1968)
- Algeria (1970)
- Bahrain (1970)
- Egypt (1973)
- United Arab Emirates (1970)
- Iraq (1972)
- Kuwait (1968)
- Libya (1968)
- Qatar (1970)
- Syria (1972).

1967 Oil Embargo

The 1967 Oil Embargo began on June 6, 1967, one day after the beginning of the Six-Day War, with a joint Arab decision to deter any countries from supporting Israel militarily. Several Middle Eastern countries eventually limited their oil shipments, some embargoing only the United States and the United Kingdom, while others placed a total ban on oil exports.

The Oil Embargo did not significantly decrease the amount of oil available in the United States or any affected European countries due mainly to a lack of solidarity and uniformity in embargoing specific countries.

The embargo was effectively ended on September 1 with the issuance of the Khartoum Resolution.

Oil Ministers' Conference

During the June 9-18 Oil Ministers' Conference in Baghdad several Arab countries issued a communiqué that two resolutions were unanimously passed:

1. "Arab oil shall be denied to and shall not be allowed to reach directly or indirectly countries committing aggression or participating in aggression on sovereignty of any Arab state or its territories or its territorial waters, particularly the Gulf of Aqaba"
2. "The involvement of any country, directly or indirectly in armed aggression against Arab states will make assets of its companies

and nationals inside the territories of Arab countries subject to the laws of war. This includes the assets of oil companies."

Invitees included the United Arab Republic, Syria, Kuwait, Libya, Saudi Arabia, Algeria, Bahrain, Abu Dhabi, and Qatar. Iraq sent copies of the Council resolution to the Embassies of Iran and Indonesia, and sought the support of Venezuela.

Oil Embargo

The Baghdad Resolution is important because Egypt broadcast claims of US aircraft support on June 6. Iraq was the first country to limit their oil shipments, embargoing the United States and the United Kingdom on June 6. Iraq, Kuwait, Algeria, Bahrain eventually embargoed the United States and the United Kingdom. Syria stopped all oil exports, rather than just embargoing specific countries in order to avoid declaring specific nations as aggressors.

The United States advocated emergency measures in OECD meetings and supported the establishment of an International Industry Advisory Board. The Advisory Board was critical in efficiently apportioning limited tanker resources and managing the distribution of the limited oil resources. This was an effective measure to negate the oil embargo as there was no consensus on what countries to embargo, and more importantly, oil shipped to a European country could then be shipped to any of the embargoed countries. Some Arab countries encouraged the oil companies to circumvent the embargo, as the Amir of Kuwait even proposed to the US ambassador that companies simply tamper with shipping manifests to allow shipment of oil to prohibited countries.

Egypt sought to bend not only international political policy but also the policies of more moderate governments. and sought to export the socialist revolution to neighboring moderate (i.e. conservative) countries. The embargo resulted in public pressure on Middle Eastern leaders to support Arab solidarity. Nasser effectively limited moderate countries' political options lest they risk a revolution.

Khartoum Resolution

The Khartoum Resolution issued on September 1 allowed the moderate oil producing nations (Kuwait, Saudi Arabia, and Libya) to resume oil exports and regain this critical source of revenue without risking disquiet or even overthrow from their more radical citizens.

In exchange, they agreed to give annual aid to "victims of Zionist aggression" namely Egypt and Jordan ($266 million and $112 million respectively). A full discussion as well as the text is available in the Khartoum Resolution article.

The oil embargo was the main reason for the formation of OAPEC which would provide a forum for the discussion of using oil politically.

1973 Oil Crisis

The 1973 oil crisis started in October 1973, when the members of Organization of Arab Petroleum Exporting Countries or the OAPEC (consisting of the Arab members of OPEC, plus Egypt, Syria and Tunisia) proclaimed an oil embargo. This was "in response to the U.S. decision to re-supply the Israeli military" during the Yom Kippur war. It lasted until March 1974.

With the U.S. actions seen as initiating the oil embargo and the long term possibility of high oil prices, disrupted supply and recession, a strong rift was created within NATO. Additionally, some European nations and Japan sought to disassociate themselves from the U.S. Middle East policy. Arab oil producers had also linked the end of the embargo with successful U.S. efforts to create peace in the Middle East, which complicated the situation.

To address these developments, the Nixon Administration began parallel negotiations with both Arab oil producers to end the embargo, and with Egypt, Syria, and Israel to arrange an Israeli pull back from the Sinai and the Golan Heights after the fighting stopped. By January 18, 1974, Secretary of State Henry Kissinger had negotiated an Israeli troop withdrawal from parts of the Sinai. The promise of a negotiated settlement between Israel and Syria was sufficient to convince Arab oil producers to lift the embargo in March 1974. By May, Israel agreed to withdraw from some parts of the Golan Heights.

Independently, the OPEAC members agreed to use their leverage over the world price setting mechanism for oil to stabilize their real incomes by raising world oil prices. This action followed several years of steep income declines after the recent failure of negotiations with the major Western oil companies earlier in the month.

Industrialized economies relied on crude oil, and OPEC was their predominant supplier. Because of the dramatic inflation experienced during this period, a popular economic theory has been that these price increases were to blame, as being suppressive of economic activity. A minority dissenting opinion questions the causal relationship described by this theory.

The targeted countries responded with a wide variety of new, and mostly permanent, initiatives to contain their further dependency. The 1973 "oil price shock", along with the 1973–1974 stock market crash, have been regarded as the first event since the Great Depression to have a persistent economic effect.

Background

Founding of OPEC: The Organization of the Petroleum Exporting Countries (OPEC), which then consisted of twelve countries, including Iran, seven Arab countries (Iraq, Kuwait, Libya, Qatar, Saudi Arabia, the United Arab Emirates), plus Venezuela, Indonesia, Nigeria, and Ecuador, had been formed at a Baghdad conference on September 14, 1960. OPEC was organized to resist pressure by the "Seven Sisters" (mostly owned by U.S., British and Dutch nationals) to reduce oil prices and payments to producing countries.

At first OPEC had operated as an informal bargaining unit for the sale of oil by resource-rich Third World nations. OPEC confined its activities to gaining a larger share of the profits generated by the Western oil companies and greater control over the members' levels of production. As a result of this and other events in the early 1970s, it began to exert its economic and political strength; the major Western oil conglomerates, as well as the importing nations, suddenly faced a unified bloc of exporters.

End of Bretton Woods

On August 15, 1971, the United States unilaterally pulled out of the Bretton Woods Accord taking the US off the Gold Exchange Standard (whereby only the value of the US dollar had been pegged to the price of gold and all other currencies were pegged to the US dollar), allowing the dollar to "float". Shortly thereafter, Britain followed, floating the pound sterling.

The industrialized nations followed suit with their respective currencies. In anticipation of the fluctuation of currencies as they stabilized against each other, the industrialized nations also increased their reserves (printing money) in amounts far greater than ever before. The result was a depreciation of the value of the US dollar, as well as the other currencies of the world. Because oil was priced in dollars, this meant that oil producers were receiving less real income for the same price. The OPEC cartel issued a joint communique stating that, from then on, they would price a barrel of oil against gold.

This led to the "Oil Shock" of the mid-seventies. In the years after 1971, OPEC was slow to readjust prices to reflect this depreciation. From 1947-1967 the price of oil in U.S. dollars had risen by less than two percent per year.

Until the Oil Shock, the price remained fairly stable versus other currencies and commodities, but suddenly became extremely volatile thereafter.

OPEC ministers had not developed the institutional mechanisms to update prices rapidly enough to keep up with changing market conditions, so their real incomes lagged for several years. The substantial price increases of 1973-74 largely caught up their incomes to Bretton Woods levels in terms of other commodities such as gold.

Yom Kippur War

On October 6, 1973, Syria and Egypt launched a surprise attack on Israel. This new round in the Arab-Israeli conflict triggered a crisis already in the making; the price of oil was going to rise. The West could not continue to increase its energy consumption 5% annually, while also paying low oil prices, and selling inflation-priced goods to the petroleum producers in the developing Third World.

This was stressed by the Shah of Iran, whose nation was the world's second-largest exporter of oil and a close ally of the United States in the Middle East at the time. "Of course [the world price of oil] is going to rise", the Shah told *The New York Times* in 1973. "Certainly! And how...; You [Western nations] increased the price of wheat you sell us by 300%, and the same for sugar and cement...; You buy our crude oil and sell it back to us, refined as petrochemicals, at a hundred times the price you've paid to us...; It's only fair that, from now on, you should pay more for oil. Let's say ten times more."

On October 12, 1973, President Richard Nixon authorized Operation Nickel Grass, an overt strategic airlift to deliver weapons and supplies to Israel, after the Soviet Union began sending arms to Syria and Egypt.

Arab Oil Embargo

On October 16, 1973, OPEC announced a decision to raise the posted price of oil by 70%, to $5.11 a barrel. The following day, oil ministers agreed to the embargo, a cut in production by five percent from September's output, and to continue to cut production over time in five percent increments until their economic and political objectives were met.

October 19, US President Richard Nixon requested Congress to appropriate $2.2 billion in emergency aid to Israel, including $1.5 billion in out-right grants. George Lenczowski notes, "Military supplies did not exhaust Nixon's eagerness to prevent Israel's collapse. ... This [$2.2B] decision triggered a collective OPEC response." Libya announced it would embargo all oil shipments to the United States. Saudi Arabia and the other OPEC states quickly followed suit, joining the embargo on October 20, 1973.

At their meeting in Kuwait the OPEC oil-producing countries, proclaimed the oil boycott that provided for curbs on their oil exports to various consumer countries and a total embargo on oil deliveries to the United States as a "principal hostile country". The embargo was thus variously extended to Western Europe and Japan.

Though United States was the initial target of the embargo, it was later expanded to the Netherlands. Price increases were also imposed. Since short term oil demand is inelastic, demand falls little when the price is raised. Thus, oil prices had to be raised dramatically to reduce demand to the new lower level of supply. Anticipating this, the market price for oil immediately rose substantially, from $3 a barrel to $12. The world financial system, which was already under pressure from the breakdown of the Bretton Woods agreement, was set on a path of recessions and high inflation that persisted until the early 1980s, with oil prices continuing to rise until 1986.

Over the long term, the oil embargo changed the nature of policy in the West towards increased exploration, energy conservation, and more restrictive monetary policy to better fight inflation.

Chronology

- January 1973—The 1973–1974 stock market crash begins, as a result of inflation pressure, the Nixon Shock and the collapsing monetary system.

- August 23, 1973—In preparation for the Yom Kippur War, Saudi King Faisal and Egyptian president Anwar Sadat meet in Riyadh and secretly negotiate an accord whereby the Arabs will use the "oil weapon" as part of the upcoming military conflict.
- October 6 - Egypt and Syria attack Israeli occupied lands in Sinai and Golan Heights on Yom Kippur, starting the fourth Arab-Israeli War.
- October 8–October 10—OPEC negotiations with major oil companies to revise the 1971 Tehran price agreement fail.
- October 12— The United States initiates Operation Nickel Grass, an overt strategic airlift operation to provide replacement weapons and supplies to Israel during the Yom Kippur War. This followed similar Soviet moves to supply the Arab side.
- October 16 - Saudi Arabia, Iran, Iraq, Abu Dhabi, Kuwait, and Qatar unilaterally raise posted prices by 17% to $3.65 per barrel and announce production cuts.
- October 17—OPEC oil ministers agree to use oil as a weapon to influence the West's support of Israel in the Yom Kippur war. They recommend an embargo against non-complying states and mandate a cut in exports.
- October 19—US President Richard Nixon requests Congress to appropriate $2.2billion in emergency aid to Israel. This decision triggered a collective Arab response. Libya proclaims an embargo on oil exports to the United States; Saudi Arabia and other Arab states follow.
- October 23–October 28—The Arab oil embargo is extended to the Netherlands.
- October 26—The Yom Kippur War ends.
- November 5—Arab producers announce a 25% output cut. A further 5% cut is threatened.
- November 23—The Arab embargo is extended to Portugal, Rhodesia, and South Africa.
- November 27—U.S. President Richard Nixon signs the Emergency Petroleum Allocation Act authorizing price, production, allocation and marketing controls.

- December 9—Arab oil ministers agree to another five percent cut for non-friendly countries for January 1974.
- December 25—Arab oil ministers cancel the five percent output cut for January. Saudi oil minister Ahmed Zaki Yamani promises a ten percent OPEC production rise.
- January 7–January 9, 1974—OPEC decides to freeze prices until April 1.
- January 18—Israel signs a withdrawal agreement to pull back to the east side of the Suez Canal.
- February 11 - United States Secretary of State Henry Kissinger unveils the Project Independence plan to make U.S. energy independent.
- February 12–February 14—Progress in Arab-Israeli disengagement brings discussion of oil strategy among the heads of state of Algeria, Egypt, Syria and Saudi Arabia.
- March 5—Israel withdraws the last of its troops from the west side of the Suez Canal.
- March 17—Arab oil ministers, with the exception of Libya, announce the end of the embargo against the United States.
- May 31—Diplomacy by Henry Kissinger produces a disengagement agreement on the Syrian front.
- December 1974—The 1973–1974 stock market crash ends.

Immediate Economic Effects

The effects of the embargo were immediate. OPEC forced the oil companies to increase payments drastically. The price of oil quadrupled by 1974 to nearly US$12 per barrel (75 US$/m^3).

This increase in the price of oil had a dramatic effect on oil exporting nations, for the countries of the Middle East who had long been dominated by the industrial powers were seen to have acquired control of a vital commodity. The traditional flow of capital reversed as the oil exporting nations accumulated vast wealth. Some of the income was dispensed in the form of aid to other underdeveloped nations whose economies had been caught between higher prices of oil and lower prices for their own export commodities and raw materials amid shrinking Western demand for their goods. Much was absorbed in massive arms purchases that exacerbated political tensions, particularly in the Middle East. This control of a vital commodity became known as the "oil weapon," which came in the form of an embargo and cutbacks in oil production from the Arab states to select industrial governments of the world to pressure Israel during the fourth Arab-Israeli War in October 1973. These target industrial governments included the United States, Great Britain, Canada, Japan, and the Netherlands.

In retrospect, the purpose of the embargo, as perceived by these target governments, was to sway their foreign policies concerning Israel towards a more pro-Arab position by threatening to cut off exports of Arab oil, and that in altering their policies the Arab states would respond by again allowing their purchase of more oil. The Arab states selected their target governments to emplace their embargo, mostly affecting the European Common Market countries and Japan with a eventual 25% oil cut in production. However, in all five cases there did not appear to be the dramatic change in policy making as envisioned by the Arab states.

In the case of the United States, scholars argue that there already existed a negotiated settlement based on equality between both parties prior to 1973. Second, Soviet involvement in the Middle East as a threat to becoming another superpower confrontation was of more concern to the United States than the oil weapon.

A third reason, the interest groups and other government agencies that were more concerned with the implications of the oil weapon held little influential power concerning foreign policy in the Arab-Israeli conflict because of Kissinger's total dominance over this process. Also within the United States concerning the economic impact at the macro level, direct correlations have been drawn between the rise in oil prices and economic recessions. "Oil price shocks", referring to disruptions in the production and distribution of oil, that result in the increase of oil prices "have been held responsible for recessions, periods of excessive inflation, reduced productivity, and lower economic growth"

The effect of the Arab embargo had a negative influence on the U.S economy through causing immediate demands to address the threats to U.S energy security. On an international level, the price increases of petroleum disrupted market systems in changing competitive positions. At the macro level, economic problems consisted of both inflationary and deflationary impacts of domestic economies.

The Arab embargo left many U.S companies searching for new ways to develop expensive oil, even in the elements of rugged terrain such as in hostile arctic environments. The problem that many of these companies faced is that finding oil and developing new oil fields usually require a time lag of 5 to 10 years between the planning process and significant oil production.

OPEC-member states in the developing world withheld the prospect of nationalization of the companies' holdings in their countries. Most notably, the Saudis acquired operating control of Aramco, fully nationalizing it in 1980 under the leadership of Ahmed Zaki Yamani. As other OPEC nations followed suit, the cartel's income soared. Saudi Arabia, awash with profits, undertook a series of ambitious five-year development plans, of which the most ambitious, begun in 1980, called for the expenditure of $250 billion. Other cartel members also undertook major economic development programs.

Meanwhile, the shock produced chaos in the West. In the United States, the retail price of a gallon of gasoline (petrol) rose from a national average of 38.5 cents in May 1973 to 55.1 cents in June 1974. State governments requested citizens not put up Christmas lights, with Oregon banning Christmas as well as commercial lighting altogether. Politicians called for a national gas rationing program. Nixon requested gasoline stations to voluntarily not sell gasoline on Saturday nights or Sundays; 90% of owners complied, which resulted in lines on weekdays.

The embargo was not uniform across Europe. Of the nine members of the European Economic Community (EEC), the Netherlands faced a complete embargo, the United Kingdom and France received almost uninterrupted supplies (having refused to allow America to use their airfields and embargoed arms and supplies to both the Arabs and the Israelis), whilst the other six faced only partial cutbacks. The UK had traditionally been an ally of Israel, and Harold Wilson's government had supported the Israelis during the Six Day War, but his successor, Ted Heath, had reversed this policy in 1970, calling for Israel to withdraw to its pre-1967 borders. The members of the EEC had been unable to achieve a common policy during the first month of the Yom Kippur War. The Community finally issued a statement on November 6, after the embargo and price rises had begun; widely seen as pro-Arab, this statement supported the Franco-British line on the war, and OPEC duly lifted its embargo from all members of the EEC. The price rises had a much greater impact in Europe than the embargo, particularly in the UK (where they combined with strikes by coal miners and railroad workers to cause an energy crisis over the winter of 1973-74, a major factor in the change of government). The UK, Germany, Italy, Switzerland, and Norway banned flying, driving and boating on Sundays. Sweden rationed gasoline and heating oil. The Netherlands imposed prison sentences for those who used more than their given ration of electricity. Ted Heath asked the British to heat only one room in their houses over the winter.

A few months later, the crisis eased. The embargo was lifted in March 1974 after negotiations at the Washington Oil Summit, but the effects of the energy crisis lingered on throughout the 1970s. The price of energy continued increasing in the following year, amid the weakening competitive position of the dollar in world markets.

Price Controls and Rationing

Government price controls further exacerbated the crisis in the United States, which limited the price of "old oil" (that already discovered) while allowing newly discovered oil to be sold at a higher price, resulting in a withdrawal of old oil from the market and the creation of artificial scarcity. The rule also discouraged alternative energies or more efficient fuels or technologies from being developed. The rule had been intended to promote oil exploration. This scarcity was dealt with by rationing of gasoline (which occurred in many countries), with motorists facing long lines at gas stations beginning in summer 1972 and increasing by summer 1973.

In 1973, U.S. President Richard Nixon named William E. Simon as the first Administrator of the Federal Energy Office, or the "Energy Czar". Simon allocated states the same amount of domestic oil for 1974 that each consumed in 1972, which worked well for states whose populations were not increasing. In states with increased populations, lines at gasoline stations were common. The American Automobile Association reported that in the last week of February 1974, 20% of American gasoline stations had no fuel at all.

In the U.S., odd-even rationing was implemented; drivers of vehicles with license plates having an odd number as the last digit (or a vanity license plate) were allowed to purchase gasoline for their cars only on odd-numbered days of the month, while drivers of vehicles with even-numbered license plates were allowed to purchase fuel only on even-numbered days. The rule did not apply on the 31st day of those months containing 31 days, or on February 29 in leap years—the latter never came into play, since the restrictions had been abolished by 1976.

In some U.S. states, a three-color flag system was used to denote gasoline availability at service stations — a green flag denoted unrationed sale of gasoline, a yellow flag denoted restricted and rationed sales, and a red flag denoted that no gasoline was available but the service station was open for other services. Additionally, coupons for gasoline rationing were ordered in 1974 and 1975 for Federal Energy Administration, but were never actually used for this crisis or the 1979 energy crisis. The rationing led to incidents of violence, after truck drivers nationwide chose to strike for two days in December 1973 because they objected to the supplies Simon had rationed for their industry. In Pennsylvania and Ohio, non-striking truckers were shot at by striking truckers, and in Arkansas, trucks of non-strikers were attacked with bombs.

America had controlled the price of natural gas since the 1950s, and with the inflation of the 1970s, the market price of natural gas was not encouraging the search for new reserves. America's natural gas reserves dwindled from 237 trillion in 1974 to 203 trillion in 1978, and the price controls were not changed despite President Gerald Ford's repeated requests to Congress.

Conservation and Reduction in Demand

To help reduce consumption, in 1974 a national maximum speed limit of 55 mph (about 88 km/h) was imposed through the Emergency Highway Energy Conservation Act. Development of the United States Strategic Petroleum Reserve began in 1975, and in 1977, the cabinet-level Department of Energy was created, followed by the National Energy Act of 1978.

Year-round daylight saving time was implemented from January 6, 1974 to February 23, 1975. The move spawned significant criticism because it forced many children to commute to school before sunrise. The pre-existing daylight-saving rules, calling for the clocks to be advanced one hour on the last Sunday in April, were restored in 1976.

The crisis also prompted a call for individuals and businesses to conserve energy, most notably a campaign by the Advertising Council using the tag line "Don't Be Fuelish." Many newspapers carried full-page advertisements that featured cut-outs which could be attached to light switches, reading "Last Out, Lights Out: Don't Be Fuelish."

By 1980, there were no longer full-size luxury cars with a 130-inch (3.3 m) wheelbase and gross weights averaging 4,500 pounds (2,041 kg). The automakers began phasing out the traditional front engine/rear wheel drive layout in favor of more efficient front engine/ front wheel drive designs.

Though not regulated by the new legislation, auto racing groups voluntarily began conserving as well. In 1974 the 24 Hours of Daytona was canceled and NASCAR reduced all race distances by 10%. The 12 Hours of Sebring race was cancelled.

In 1976, the U.S. Congress created the Weatherization Assistance Program to help low-income homeowners and renters deal with rising heating costs by reducing their demand through advanced insulation.

Secondary Effects

Various secondary effects occurred, notably toilet paper panics in Japan and the United States; these were unfounded panics which became self-fulfilling prophesies, and are classic examples of the Thomas theorem.

Price rises and unfounded rumors of a toilet paper shortage – based on oil being used in paper manufacturing – caused a panic and hoarding of toilet paper in late October and early November in Osaka and Kobe, among other cities. In the US, Johnny Carson inadvertently caused a three-week panic when, on December 19, 1973, he read a news item regarding the US government falling behind on bids for toilet paper and quipping that the nation faced a toilet paper shortage on the *Tonight Show*.

Search for Alternatives

The energy crisis led to greater interest in renewable energy and spurred research in solar power and wind power. It also led to greater pressure to exploit North American oil sources, and increased the West's dependence on coal and nuclear power. This included increased interest in mass transit.

In Australia, heating oil ceased being considered an appropriate winter heating fuel. This often meant that a lot of oil-fired room heaters that were popular from the late-1950s to the early-1970s were considered outdated. Gas-conversion kits that let the heaters burn natural gas or propane were introduced.

For the handful of industrialized nations that were net energy exporters, the effects of the oil crisis were very different. In Canada the industrial east suffered many of the same problems of the United States. In oil rich Alberta, however, there was a sudden and massive influx of money that quickly made it the richest province in the country. The federal government attempted to correct this imbalance through the creation of the government-owned Petro-Canada and later the National Energy Program. These efforts produced a great deal of anger in the west producing a sentiment of alienation that has remained a central element of Canadian politics to this day. Overall the oil embargo had a sharply negative effect on the Canadian economy. The economic malaise in the United States easily crossed the border and increases in unemployment, and stagflation hit Canada as hard as the United States despite Canadian fuel reserves.

The Brazilian government implemented a very large project called "Proálcool" (pro-alcohol) that mixed ethanol with gasoline for automotive fuel.

To supplement Israel's over-taxed power grid, Harry Zvi Tabor, the father of Israel's solar industry, developed the prototype for a solar water heater now used in over 90% of Israeli homes.

Macroeconomic Effects

The 1973 oil crisis was a major factor in Japan's economy shifting from oil-intensive industries, and resulted in huge Japanese investments in industries such as electronics. The Japanese auto makers also took advantage of this embargo. After they realized what fuel costs were in the United States, they started producing small, more fuel efficient models, which began selling as an alternative to "gas-guzzling" American vehicles of the time. This triggered a drop in American auto sales that lasted into the 1980s.

The Western nations' central banks decided to sharply cut interest rates to encourage growth, deciding that inflation was a secondary concern. Although this was the orthodox macroeconomic prescription at the time, the resulting stagflation surprised economists and central bankers, and the policy is now considered by some to have deepened and lengthened the adverse effects of the embargo.

Long-term effects of the embargo are still felt. Many in the public remain suspicious of oil companies, believing they profiteered, or even colluded with OPEC. In 1974, seven of the fifteen top Fortune 500 companies were oil companies.

Effects on International Relations

The Cold War policies of the Nixon administration also suffered a major blow in the aftermath of the oil embargo. They had focused on China and the Soviet Union, but the latent challenge to U.S. hegemony coming from the Third World became evident. U.S. power was under attack even in Latin America.

The oil embargo was announced roughly one month after a right-wing military coup in Chile led by General Augusto Pinochet Chilean coup of 1973 toppled socialist president Salvador Allende on September 11, 1973.

The United States' subsequent assistance to this government did little to curb the activities of socialist guerrillas in the region. The response of the Nixon administration was to propose doubling of the amount of military arms sold by the United States. As a consequence, a Latin American bloc was organized and financed in part by Venezuela and its oil revenues, which quadrupled between 1970 and 1975.

In addition, Western Europe and Japan began switching from pro-Israel to more pro-Arab policies. This change further strained the Western alliance system, for the United States, which imported only 12% of its oil from the Middle East (compared with 80% for the Europeans and over 90% for Japan), remained staunchly committed to backing Israel. The percentage of U.S. oil which comes from the nations bordering the Persian Gulf has remained steady over the years, with a figure of a little more than 10% in 2008.

Although historically having no connections to the Middle East, Japan was the most heavily dependent on its oil from this region, making up 71% of its imported oil from the Middle East in 1970. However, on November 7, 1973, the Saudi and Kuwaiti governments declared Japan a "nonfriendly" country directed towards changing its policy of noninvolvement in the Arab-Israeli conflict, placing a 5 percent production cut in December to Japan. The December production cut to the Japanese government caused somewhat of a panic, where on November 22 Japan issued a statement "asserting that Israel should withdraw from all of the 1967 territories, advocating Palestinian self-determination, and threatening to reconsider its policy toward Israel if Israel refused to accept these preconditions" By December 25, Japan was considered a friendly state.

With the oil embargo in place, the industrial governments of the world in some way altered their foreign policy regarding the Arab-Israeli conflict and after the use of the Arab oil weapon. These included European countries such as the UK who decided to refuse to allow the United States to use British bases in the UK and in Cyprus to airlift resupplies to Israel along with the rest of the members of the European Community.

It also included the Japanese restatement on November 22, to "reconsider" their relations with Israel if Israel did not acknowledge their avocations to return to their pre-1967 territorial state, although this was never acted upon. Canada shifted towards a more pro-Arab position after displeasure was expressed by many Arab governments towards Canada's Middle Eastern position as one of being mostly neutral. "On the other hand, after the embargo the Canadian government moved quickly indeed toward the Arab position, despite its low dependence on Middle Eastern oil"

A year after the start of the 1973 oil embargo, the nonaligned bloc in the United Nations passed a resolution demanding the creation of a "new international economic order" in which resources, trade, and markets would be distributed more equitably, with the local populations of nations within the global South receiving a greater share of benefits derived from the exploitation of southern resources, and greater respect for the right to self-directed development in the South be afforded by the North.

In the post-Cold War era, Israel continues to serve the U.S.A. as a strategically important non-NATO ally in the Middle East. According to the American military journalist and commentator William M. Arkin in his book *Code Names*, the U.S. has prepositioned munitions, vehicles, and military equipment, and even a 500-bed hospital for use by US Marines, Special Forces, and Air Force fighter and bomber aircraft in a wartime contingency at least six sites in Israel. Late Republican Senator Jesse Helms used to call Israel "America's aircraft carrier in the Middle East", when explaining why the US viewed Israel as such a strategic ally, saying that the military foothold in the region offered by the Jewish State alone justified the military aid that the US grants Israel every year. Israel is not the only country in the Middle East to host US military bases, though. There are American military facilities in Egypt, Jordan, Saudi Arabia, Oman, and the Persian Gulf states of Kuwait, Bahrain (headquarters of the United States Fifth Fleet), and Qatar.

Decline of OPEC

Since 1973, OPEC failed to hold on to its preeminent position, and by 1981, its production was surpassed by that of other countries. Additionally, its own member nations were divided among themselves. Saudi Arabia, trying to gain back market share, increased production and caused downward pressure on prices, making high-cost oil production facilities less profitable or even unprofitable.

The world price of oil, which had reached a peak in 1979 during the 1979 energy crisis, at more than US$80 per barrel, decreased during the early 1980s to US$38 per barrel (239 US$/m^3). In real prices, oil briefly fell back to pre-1973 levels. Overall, the reduction in price was a windfall for the oil-consuming nations: United States, Japan, Europe, and especially the Third World.

Part of the decline in prices and economic and geopolitical power of OPEC comes from the move away from oil consumption to alternate energy sources. OPEC had relied on the famously limited price inelasticity of oil demand to maintain high consumption but had underestimated the extent to which other sources of supply would become profitable as the price increased.

Electricity generation from nuclear power and natural gas, home heating from natural gas and ethanol blended gasoline all reduced the demand for oil. At the same time, the drop in prices represented a serious problem for oil-producing countries in northern Europe and the Persian Gulf region. For a handful of heavily populated, impoverished countries, whose economies were largely dependent on oil — including Mexico, Nigeria, Algeria, and Libya — governments and business leaders failed to prepare for a market reversal, the price drop placed them in wrenching, sometimes desperate situations.

When reduced demand and over-production produced a glut on the world market in the mid-1980s, oil prices plummeted and the cartel lost its unity. Oil exporters such as Mexico, Nigeria, and

Venezuela, whose economies had expanded in the 1970s, were plunged into near-bankruptcy, and even Saudi Arabian economic power was significantly weakened.

The divisions within OPEC made subsequent concerted action more difficult. Nevertheless, the 1973 oil shock provided dramatic evidence of the potential power of Third World resource suppliers in dealing with the developed world. The vast reserves of the leading Middle East producers guaranteed the region its strategic importance, but the politics of oil still proves dangerous for all concerned to this day.

Long-term Effects

Prior to the embargo, the geo-political competition between the Soviet Union and the United States, in combination with low oil prices that hindered the necessity and feasibility for the West to seek alternative energy sources, presented the Arab States with financial security, moderate economic growth, and disproportionate international bargaining power. Following the embargo, higher oil prices instigated new avenues for energy exploration or expansion including Alaska, the North Sea, the Caspian Sea, and Caucasus.

Soviet Reaction

Prior to the ascendancy of Anwar Sadat to president of Egypt in 1970, the Middle East had been an important arena in the global superpower competition, most lucidly displayed in the arms sales and cooperation between the American and Soviet governments with Israel, Saudi Arabia, and Iran allied to The United States, and Egypt, Syria, and Iraq allied with the Soviet Union.

Although none of these states entered into any formal alliances comparative to the North Atlantic Treaty Organization, they did benefit greatly from the geo-political competition in the region and vacillations in alignment often resulted in greater gains of assistance. This competitive environment, beneficial to the regional states involved, was mitigated sharply after 1970. Sadat's dismissal of Soviet specialists in Egypt and the dramatic price increases in hydrocarbons hardened relations with all of the Middle East and created new opportunities for the export of Soviet oil.

Exploration in the Caspian Basin and Siberia became more cost effective. Former cooperation evolved into a far more adversarial relationship as the Soviet Union increased oil production and export (by 1980 the Soviet Union was the world's largest producer of oil) to take advantage of the supply problems in the West created by OPEC's production reductions. This growing economic competition turned into genuine fears of military aggression after the 1979 Soviet invasion of Afghanistan, leaving the Persian Gulf states to look to the United States for the type of security guarantees against Soviet military action in the Persian Gulf that the Israelis had exclusively received only a decade earlier.

Growing Security Concerns

The Soviet invasion of Afghanistan was only part of the growing security destabilization in the Middle East, most obviously seen in the increased sale of American weapons, technology, and outright military presence. Saudi Arabia and Iran became increasingly dependent on bi-lateral American security assurances to combat both external and internal threats, including increased military competition between these states because of the increased oil revenues.

Both states were seemingly competing for preeminence in the Persian Gulf and using increased revenues on disproportionately powerful military forces. By 1979, Saudi weapon purchases from the United States were in excess of five times the amount that Israel was purchasing annually.

Following the failure of the Shah during January 1979 to maintain control of Iran, the Saudis were forced to deal with the prospect of internal destabilization via Islamic fundamentalism, a reality which would quickly be revealed in the seizure of the Grand Mosque in Mecca by Wahhabi extremists during November and a Shia revolt in al-Hasa during December.

Growing fears about eventual Western energy independence, various security threats, and the absence of a Western rival in the geo-political competition over the Middle-East led the Arab states in a more dependent relationship with the West. This is most explicit in Saudi Arabia's consistent policy of price and production moderation in an effort to reduce the chances of Western alienation and the opportunity costs for alternative energy production. The exchange for Western moderation in Arab-Israeli affairs ultimately led to a reshaping of the Middle-Eastern geo-political landscape that was significantly less advantageous than prior to 1973.

Impact on Motor Industry

West Europe

The motor industry was one of Western Europe's most affected industries in the wake of the 1973 oil crisis.

After the Second World War most West European countries applied heavy taxes to motor fuel because it was imported, and as a result most cars made in Europe were small and economical. However by the late sixties as wealth increased car sizes were rising despite heavy fuel taxes, although some of the more upmarket brands were building cars that could take lead-free fuel, and there were still a number of "economy" cars in production at this time.

But the oil crisis gradually saw many West European car buyers move away from larger, less economical cars. The most notable result of this transition in the car market was the rise in popularity of compact hatchbacks. The only notable small hatchbacks built in Western Europe at the time of the oil crisis were the Peugeot 104, Renault 5 and Fiat 127. By the end of the decade, the market had massively expanded with the introduction of the Ford Fiesta, Opel Kadett (sold as the Vauxhall Astra in Great Britain), Chrysler Sunbeam and Citroën Visa.

Buyers looking for larger cars were increasingly drawn to medium sized hatchbacks that were virtually unknown in Europe in 1973, but by the end of the decade were gradually replacing saloons as the mainstay of this sector. Between 1973 and 1980, the following medium sized hatchbacks were launched across Europe: the Chrysler/Simca Horizon, Fiat Ritmo (Strada in the UK), Ford Escort MK3, Renault 14, Volvo 340 / 360, Opel Kadett and Volkswagen Golf. These cars offered new standard of fuel economy, which were much needed in the aftermath of the oil crisis.

The new cars launched in the wake of the oil crisis were considerably more economical than the traditional saloons they were taking the place of, and even attracted a considerable number of buyers who would have otherwise chosen cars in the next sector. Their success continued into the 1980s and by the later part of the decade, some 15 years after the oil crisis, hatchbacks almost monopolised most European small and medium car markets, and had gained a substantial share of the large family car market.

US

As in Western Europe, U.S. automakers were significantly impacted by the 1973 oil embargo and energy crisis. Before the energy crisis, large, heavy, and powerful cars were the standard in the U.S. By 1971, the standard engine in a Chevrolet Caprice was a 400-cubic inch (6.5 liter) V8. The wheelbase of this car was 121.5 inches (3,090 mm), and *Motor Trend*'s 1972 road test of the similar Chevrolet Impala logged no more than 15 miles per gallon on the highway.

After the energy crisis, however, gasoline cost more and reduced the demand for large cars. The Toyota Corona, the Toyota Corolla, the Datsun B210, the Datsun 510, the Honda Civic, the Mitsubishi Galant (a captive import from Chrysler sold as the Dodge Colt), the Subaru DL, and later the Honda Accord all had four cylinder engines that were more fuel efficient in comparison to the typical V8 and six cylinder engines found in North American vehicles. From Europe, the Volkswagen Beetle, the Volkswagen Fastback, the Renault 8, the Renault LeCar, and the Fiat Brava were also offered. As buyers began exchanging large cars for the smaller imported ones, Detroit responded with the Ford Pinto, the Ford Maverick, the Chevrolet Vega, the Chevrolet Nova, the Plymouth Valliant, and the Plymouth Volaré.

Some buyers lamented the small size of the first compacts that came from Japan, and both Toyota and Nissan (known as Datsun during the 1970s) introduced larger cars called the Toyota Corona Mark II, replaced by the Toyota Cressida, the Mazda 616, and Datsun 810 which gave buyers increased passenger space and some luxury amenities, such as air conditioning, power steering, AM-FM radios, and even power windows and central locking without increasing the price of the vehicle. These larger compacts were at the very limit of Japanese government regulations concerning size and engine displacement so that they could still be affordable in the Japanese Domestic Market, yet offer export buyers larger cars that sacrificed fuel economy for passenger accommodation and a higher price. Toyota also sold the Toyota Crown from 1965 to 1974 with very limited amount of sales.

Compact trucks were also introduced to the USA, with the Toyota Hilux and the Datsun Truck, followed by the Mazda Truck also sold as the Ford Courier, with Isuzu selling their compact truck as the Chevrolet LUV.

An increase in imported cars into North America forced the Big Three (General Motors, Ford, and Chrysler) to introduce smaller and fuel-efficient models for domestic sales.

The Dodge Omni / Plymouth Horizon from Chrysler, the Ford Fiesta, and the Chevrolet Chevette all had four-cylinder engines and room for at least four passengers by the late '70s. By 1985, the average American vehicle received 17.4 miles per gallon, compared to 13.5 miles per gallon in 1970. The improvements stayed even though the price of a barrel of oil remained constant at $12 from 1974 to 1979.

While at the same time these new imports were major inroads in the American market, sales of large sedans for most makes (except Chrysler products) recovered within two model years of the 73' Oil Crisis. Sales of models such as the Cadillac DeVille, Buick Electra, Oldsmobile 98, Lincoln Continental, Mercury Marquis, and various other luxury oriented sedans became popular again in the mid-70s.

The only full-size models to see permanent reductions in sales were the lower price models; such as the Chevrolet Impala, and Ford Galaxie 500. At the same time, slightly smaller, if not entirely more fuel efficient mid-size models such as the Oldsmobile Cutlass, Chevrolet Monte Carlo, Ford Thunderbird and various other models sold well.

This led to the somewhat odd juxtaposition of small economical imports introducing themselves as major elements of the market, while at the same time heavy, expensive, largely impractical vehicles (with 7 mpg; Lincoln sold 80,321 Mark Vs in 77) selling alongside the new imports in equally impressive numbers. In 1976; Toyota, with an average weight around 2,100 lbs sold 346,920 cars in the United States, while Cadillac with an average weight around 5,000 lbs sold 309,139 cars.

Federal safety standards, such as NHTSA Federal Motor Vehicle Safety Standard 215 (pertaining to safety bumpers), and compacts like the 1974 Mustang II were a prelude to the DOT "downsize" revision of vehicle categories. By 1977, GM's full-sized cars reflected on the 1973 oil crisis and preceded later DOT downsizing. By 1979, virtually all the big "full size" American cars were "downsized", featuring smaller engines and smaller dimensions outside. Chrysler Corporation ended production of their full-sized luxury sedans at the end of the 1981 model year, moving instead to a full front wheel drive lineup for 1982 (except for the M-body Dodge Diplomat/Plymouth

Gran Fury and Chrysler New Yorker Fifth Avenue sedans). It has been suggested that if mass production of overdrive transmissions had been introduced, there would not have actually been any vehicle downsizing. But since this has actually happened it turns out not to be true.

1990 Oil Price Shock

The 1990 oil price spike occurred in response to the Iraqi invasion of Kuwait on August 2, 1990. Lasting only 9 months, the price shock was less extreme and of shorter duration than the previous oil crises of 1973 and 1979-1980, yet the rise in prices is widely believed to have been a significant factor in the recession of the early 1990s.

Average monthly prices of oil rose from $17 per barrel in July to $36 per barrel in August. As the U.S.-led coalition experienced military success against Iraqi forces, concerns about long-term supply shortages eased and prices began to fall.

Iraqi Invasion of Kuwait and Ensuing Economic Effects

On August 6, 1990, The Republic of Iraq invaded the State of Kuwait, leading to a 7-month occupation of Kuwait and an eventual U.S.-led military intervention.

While Iraq officially claimed Kuwait was stealing its oil via slant drilling, its true motives are more complicated and less clear. At the time of the invasion, Iraq owed Kuwait $14 billion of outstanding debt that Kuwait had loaned it during the Iraq-Iran war.

In addition, Iraq felt Kuwait was overproducing oil, lowering prices and hurting Iraqi oil profits in a time of financial stress.

In the buildup to the invasion, Iraq and Kuwait had been producing 4.3 million barrels (680,000 m^3) of oil a day. This potential loss, coupled with threats to Saudi Arabian oil production, led to a rise in prices from $21 per barrel at the end of July to $28 per barrel on August 6. On the heels of the invasion, prices rose to a peak of $46 per barrel in mid-October.

The United States' rapid intervention and subsequent military success helped to mitigate the potential risk to future oil supplies, thereby calming the market and restoring confidence. After only three quarters, or 9 months, the spike had subsided.

US Policy Response

The U.S. Federal Reserve's monetary tightening in 1988 targeted the rapid inflation of the 1980s. By increasing the federal funds rate and lowering growth expectations, the Fed hoped to slow and eventually reduce inflationary pressures, creating greater price stability. The August 6 invasion was seen as a direct threat to the price stability the Fed sought. In fact, the Council of Economic Advisors published a consensus estimate that a one-year, 50 percent increase in the price of oil could temporarily raise the price level of the economy by 1 percent and potentially lower real output by the same amount.

Despite the potential for inflation, the U.S. Fed and central banks around the globe decided it would not be necessary to raise interest rates to counteract the rise in oil prices. Rather, the U.S. Federal Reserve decided to maintain interest rates as if the oil price spike were not occurring.

This decision to refrain from action stemmed from confidence in the future success of Desert Storm to protect major oil-producing facilities in the Middle East and a will to maintain the long-term credibility of economy policy that had been built up during the 1980s.

To avoid being accused of inaction in the face of potential economic turbulence, the U.S. revised the Gramm-Rudman-Hollings Balanced Budget Act. Initially, the act prohibited the U.S. from changing budget deficit targets even in the event of a negative shock to the economy. When oil prices rose, revision of this act allowed the U.S. government to adjust its budget for changes in the economy, further mitigating the risk of rising prices. The result was a peak in prices at $46 per barrel in mid-October, followed by a steady decline in prices until 1994.

Supply Shock

A supply shock is an event that suddenly changes the price of a commodity or service. It may be caused by a sudden increase or decrease in the supply of a particular good. This sudden change affects the equilibrium price. A negative supply shock (sudden supply decrease) will raise prices and shift the aggregate supply curve to the left. A negative supply shock can cause stagflation due to a combination of raising prices and falling output. A positive supply shock (an increase in supply) will lower the price of said good and shift the aggregate supply curve to the right. A positive supply shock could be an advance in technology (a technology shock) which makes production more efficient, thus increasing output. An example of a negative supply shock is the increase in oil prices during the 1973 energy crisis.

Technical Analysis

The diagram to the right demonstrates a negative supply shock; The initial position is at point A, producing Y_1 quantity, at P_1 prices. Then there is a supply shock, this has an adverse effect on aggregate supply, the supply curve shifts left (from AS_1 to AS_2), while the demand curve stays in the same position. The intersection of the supply and demand curves has now moved and the equilibrium is now point B, quantity has been reduced to Y_2, while prices have been increased to P_2.

1979 Energy Crisis

The 1979 (or second) oil crisis in the United States occurred in the wake of the Iranian Revolution. Amid massive protests, the Shah of Iran, Mohammad Reza Pahlavi, fled his country in early 1979 and the Ayatollah Khomeini soon became the new leader of Iran. Protests severely disrupted the Iranian oil sector, with production being greatly curtailed and exports suspended.

When oil exports were later resumed under the new regime, they were inconsistent and at a lower volume, which pushed prices up. Saudi Arabia and other OPEC nations, under the presidency of Dr. Mana Alotaiba increased production to offset the decline, and the overall loss in production was about 4 percent. However, a widespread panic resulted, added to by the decision of U.S. President Jimmy Carter to order the cessation of Iranian imports, driving the price far higher than would be expected under normal circumstances. In April of the same year, President Carter began a phased deregulation of oil prices. At the time, the average price of crude oil was $15.85 per barrel (42 US gallons (160 L)). Deregulating domestic oil price controls allowed U.S. oil output to rise sharply from the Prudhoe Bay fields, although oil imports fell sharply. Long lines once again appeared at gas stations and convenience stores, just as they did in 1973.

In 1980, following the outbreak of the Iran–Iraq War, oil production in Iran nearly stopped, and Iraq's oil production was severely cut as well. After 1980, oil prices began a 20-year decline down to a 60 percent price drop in the 1990s. Oil exporters such as Mexico, Nigeria, and Venezuela expanded production ; USSR became the first world producer, and North Sea and Alaskan oil flooded onto the market.

Iran

In November 1978, a strike by 37,000 workers at Iran's nationalized oil refineries initially reduced production from 6 million barrels (950,000 m^3) per day to about 1.5 million barrels (240,000 m^3). Foreign workers (including skilled oil workers) fled the country. On January 16, 1979, Shah of Iran, Mohammad Reza Pahlavi and his wife left Iran at the behest of Prime Minister Shapour Bakhtiar (a long time opposition leader himself), who sought to calm down the situation.

Effects

Other OPEC members: The rise in oil price benefited other OPEC members, which made record profits.

United States

The oil crisis had mixed effects in the United States, due to some parts of the country being oil-producing regions and other parts being oil-consuming regions. Richard Nixon had imposed price controls on domestic oil, which had helped cause shortages that led to gasoline lines during the 1973 Oil Crisis. Gasoline controls were repealed, but controls on domestic US oil remained.

The Jimmy Carter administration began a phased deregulation of oil prices on April 5, 1979, when the average price of crude oil was US$15.85 per barrel (42 US gallons (160 L)). Starting with the Iranian revolution, the price of crude oil rose to $39.50 per barrel over the next 12 months (its all time highest real price until March 7, 2008.) Deregulating domestic oil price controls allowed domestic U.S. oil output to rise sharply from the large Prudhoe Bay fields, while oil imports fell sharply.

Due to memories of oil shortage in 1973, motorists soon began panic buying, and long lines appeared at gas stations, as they had six years earlier during the 1973 oil crisis.

As the average vehicle of the time consumed between two to three liters (about 0.5-0.8 gallons) of gasoline (petrol) an hour while idling, it was estimated that Americans wasted up to 150,000 barrels (24,000 m^3) of oil per day idling their engines in the lines at gas stations.

During the period, many people believed the oil companies artificially created oil shortages to drive up prices, rather than factors beyond human control or the US' own price controls.

The amount of oil sold in the United States in 1979 was only 3.5 percent less than the record set for oil sold the year previously.

Many politicians proposed gas rationing; one such proponent was Harry Hughes, Governor of Maryland, who proposed odd-even rationing (only people with an odd-numbered license plate could purchase gas on an odd-numbered day), as was used during the 1973 Oil Crisis.

Several states actually implemented odd-even gas rationing, including Pennsylvania, New York, New Jersey, and Texas. Coupons for gasoline rationing were printed but were never actually used during the 1979 crisis.

On July 15, 1979, President Jimmy Carter outlined his plans to reduce oil imports and improve energy efficiency in his "Crisis of Confidence" speech (sometimes known as the "malaise" speech). It is often said that during the speech, Carter wore a cardigan (he actually wore a blue suit) and encouraged citizens to do what they could to reduce their use of energy. He had already installed solar power panels on the roof of the White House and a wood-burning stove in the living quarters. However, the panels were removed in 1986, reportedly for roof maintenance, during the administration of his successor, Ronald Reagan.

Carter's speech argued the oil crisis was "the moral equivalent of war". Several months later, in January 1980, Carter issued the Carter Doctrine, which declared that any interference with U.S. oil interests in the Persian Gulf would be considered an attack on the vital interests of the United States. Additionally, as part of his administration's efforts at deregulation, Carter proposed removing price controls that had been imposed in the administration of Richard Nixon before the 1973 crisis. Carter agreed to remove price controls in phases; they were finally dismantled in 1981 under Reagan. Carter also said he would impose a windfall profit tax on oil companies. While the regulated price of domestic oil was kept to $6 a barrel, the world market price was $30.

In 1980, the U.S. Government established the Synthetic Fuels Corporation to produce an alternative to imported fossil fuels.

Oil Patch

When the price of West Texas Intermediate crude oil increased 250 percent between 1978 and 1980, the oil-producing areas of Texas, Oklahoma, Louisiana, Colorado, Wyoming, and Alaska began experiencing an economic boom and population inflows.

Automobile Fuel Economy

At the same time, Detroit's then-Big Three automakers (Ford, Chrysler, GM) were marketing downsized full-sized automobiles like the Chevrolet Caprice, the Ford LTD Crown Victoria and the Dodge St. Regis which met the CAFE fuel economy mandates passed in 1978. Detroit's response to the growing popularity of imported compacts like the Toyota Corolla and the Volkswagen Rabbit were the Chevrolet Citation, and the Ford Fairmont; Ford replaced the Ford Pinto with the Ford Escort and Chrysler, on the verge of bankruptcy, introduced the Dodge Aries K. GM was having unfavorable market reactions to the Citation, and introduced the Chevrolet Corsica and Chevrolet Beretta in 1987 which did sell better.

GM also replaced the Chevrolet Monza, introducing the 1982 Chevrolet Cavalier which was better received. Ford experienced a similar market rejection of the Fairmont, and introduced the front wheel drive Ford Tempo in 1984.

Detroit was not well prepared for the sudden rise in fuel prices, and imported brands were now more widely available in North America and had developed a loyal customer base. Many imported brands utilized fuel saving technologies such as fuel injection and multi-valve engines over the common use of carburetors. GM's Cadillac division experimented with their V8-6-4 power plant (the ancestor of the modern-day Active Fuel Management and/or variable displacement), which was a market failure. Nonetheless, overall fuel economy increased, which was one factor leading to the subsequent 1980s oil glut.

1980s Oil Glut

The 1980s oil glut was a serious surplus of crude oil caused by falling demand following the 1970s Energy Crisis. The world price of oil, which had peaked in 1980 at over US$35 per barrel ($93 per barrel today), fell in 1986 from $27 to below $10 ($54 to $20 today). The glut began in the early 1980s as a result of slowed economic activity in industrial countries (due to the crises of the 1970s, especially in 1973 and 1979) and the energy conservation spurred by high fuel prices. The inflation adjusted real 2004 dollar value of oil fell from an average of $78.2 in 1981 to an average of $26.8 per barrel in 1986.

In June 1981, *The New York Times* stated an "Oil glut! ... is here" and *Time Magazine* stated: "the world temporarily floats in a glut of oil," though the next week an article in *The New York Times* warned that the word "glut" was misleading, and that in reality, while temporary surpluses had brought down prices somewhat, prices were still well above pre-energy crisis levels.

This sentiment was echoed in November 1981, when the CEO of Exxon Corp also characterized the glut as a temporary surplus, and that the word "glut" was an example of "our American penchant for exaggerated language." He wrote that the main cause of the glut was declining consumption.

In the United States, Europe and Japan, oil consumption had fallen 13% from 1979 to 1981, due to "in part, in reaction to the very large increases in oil prices by the Organization of Petroleum Exporting Countries and other oil exporters," continuing a trend begun during the 1973 price increases.

After 1980, reduced demand and overproduction produced a glut on the world market, causing a six-year-long decline in oil prices culminating with a 46 percent price drop in 1986.

Background

The 1973 oil crisis and the 1979 oil crisis turned oil from a cheap to a very expensive energy source. During the 1973 energy crisis, the price of oil quadrupled. Oil never returned to pre-1973 levels, either in real or nominal terms, even during the 1980s glut.

The nominal price continued its slow increase after the crisis ended. Six years later, the price more than doubled during the 1979 energy crisis. OPEC and Saudi Arabia artificially raised the price of oil several times in 1979 and 1980. Also during this time, several OPEC members significantly lowered their production levels, the Iran hostage crisis occurred, and the Iran–Iraq War began. There was fear that the world's oil market supply was tenuous, causing the price of oil to escalate and that OPEC would dictate very high prices in a shortage.

Production

Non-OPEC: During the 1980s, non-OPEC production increased worldwide.

US

In April 1979, Jimmy Carter signed an executive order which was to remove market controls from petroleum products by October 1981, so that prices would be wholly determined by the free market. Ronald Reagan signed an executive order on January 28, 1981 which enacted this reform immediately, allowing the free market to adjust oil prices in the US. This ended the withdrawal of old oil from the market and artificial scarcity, encouraging increased oil production. The US Oil Windfall profits tax was lowered in August 1981 and removed in 1988, ending disincentives to US oil producers. Additionally, the Alaskan Prudhoe Bay Oil Field entered peak production, supplying the US West Coast with up to 2 million bpd of crude oil.

OPEC

From 1980 to 1986, OPEC decreased oil production several times and nearly in half to maintain oil's high prices. However, it failed to hold on to its preeminent position, and by 1981, its production was surpassed by Non-OPEC countries. OPEC had seen its share of the world market drop to less than a third in 1985, from nearly half during the 1970s. In Feb 1982, the *Boston Globe* reported that OPEC's production, which had previously peaked in 1977, was at its lowest level since 1969. Non-OPEC nations were at that time supplying most of the West's imports. OPEC's membership began to have divided opinions over what actions to take. In September 1985, Saudi Arabia tried to gain market share by increasing production, creating a "huge surplus that angered many of their colleagues in OPEC". High-cost oil production facilities became less or even not profitable.

US Imports

The US imported 28 percent of its oil in 1982 and 1983, down from 46.5 percent in 1977, due to lower consumption. Reliance on Middle East sources dwindled even further as Britain, Mexico, Nigeria and Norway joined Canada in the forefront of American suppliers.

Imported crude oil from Libya was banned in the United States on March 10, 1982.

Reduced Demand

OPEC had relied on the price elasticity of demand of oil to maintain high consumption, but underestimated the extent to which other sources of supply would become profitable as prices increased. Electricity generation from nuclear power and natural gas; home heating from natural gas; and ethanol blended gasoline all reduced the demand for oil. New passenger car fuel economy rose from 17 mpg in 1978 to more than 22 mpg in 1982, an increase of more than 30 percent.

Impact

The 1986 oil price collapse benefited oil-consuming countries such as the United States, Japan, Europe, and Third World nations, but represented a serious loss in revenue for oil-producing countries in northern Europe, the Soviet Union, and OPEC. In 1981, before the brunt of the glut, *Time Magazine* wrote that in general, "A glut of crude causes tighter development budgets" in some oil-exporting nations. In a handful of heavily populated impoverished countries whose economies were largely dependent on oil production — including Mexico, Nigeria, Algeria, and Libya — government and business leaders failed to prepare for a market reversal. With the drop in oil prices, OPEC lost its unity. Oil exporters such as Mexico, Nigeria, and Venezuela, whose economies had expanded in the 1970s, were plunged into near-bankruptcy. Even Saudi Arabian economic power was significantly weakened. Iraq had fought a long and costly war against Iran, and had particularly weak revenues. It was upset by Kuwait contributing to the glut and allegedly pumping oil from the Rumaila field below their common border. Iraq invaded Kuwait territory in 1990, planning to increase reserves and revenues and cancel the debt, resulting in the first Gulf War. The USSR had become a major oil producer before the glut. The drop of oil prices contributed to the nation's final collapse.

In the US, domestic exploration declined dramatically, and the number of active drilling rigs was nearly halved in 1982." Oil producers held back on the search for new oilfields for fear of losing on their investments. In May 2007, companies like ExxonMobil were not making nearly the investment in finding new oil today that they did in 1981. Cities that had experienced tremendous growth when oil prices were high, most notably Houston and New Orleans, endured severe local recessions as prices collapsed.

Price of Petroleum

The price of petroleum as quoted in news generally refers to the spot price per barrel (159 liters) of either WTI/light crude as traded on the New York Mercantile Exchange (NYMEX) for delivery at Cushing, Oklahoma, or of Brent as traded on the Intercontinental Exchange (ICE, into which the International Petroleum Exchange has been incorporated) for delivery at Sullom Voe.

The price of a barrel of oil is highly dependent on both its grade, determined by factors such as its specific gravity or API and its sulphur content, and its location. Other important benchmarks include Dubai, Tapis, and the OPEC basket.

The Energy Information Administration (EIA) uses the imported refiner acquisition cost, the weighted average cost of all oil imported into the US, as its "world oil price". The demand for oil is highly dependent on global macroeconomic conditions. According to the International Energy Agency, high oil prices generally have a large negative impact on the global economic growth. The Organization of the Petroleum Exporting Countries (OPEC) was formed in 1960 to try and counter the oil companies cartel, which had been controlling posted prices since the so-called 1927 Red Line Agreement and 1928 Achnacarry Agreement, and had achieved a high level of price stability until 1972.

The price of oil underwent a significant decrease after the record peak of US$145 it reached in July 2008. On December 23, 2008, WTI crude oil spot price fell to US$30.28 a barrel, the lowest since the financial crisis of 2007–2010 began, and traded at between US$35 a barrel and US$82 a barrel in 2009. On 31 January 2011, the Brent price hit $100 a barrel for the first time since October 2008, on concerns about the political unrest in Egypt.

Recent Price History

A recent low point was reached in January 1999 of $17 (all prices are in US$ per barrel), after increased oil production from Iraq coincided with the Asian Financial Crisis, which reduced demand. Prices then increased rapidly, more than doubling by September 2000 to $35, then fell until the end of 2001 before steadily increasing, reaching $40–50 by September 2004. In October 2004, light crude futures contracts on the NYMEX for November delivery exceeded $53 and for December delivery exceeded $55. Crude oil prices surged to a record high above $60 in June 2005, sustaining a rally built on strong demand for gasoline and diesel and on concerns about refiners' ability to keep up. This trend continued into early August 2005, as NYMEX crude oil futures contracts surged past $65 as consumers kept up the demand for gasoline despite its high price.

Crude oil futures peaked at a close of over $77 in July 2006, and in December 2006 at about $63. That is just about where they began the year 2006. In September 2007, US crude (WTI) crossed $80. Multiple factors caused this high price. OPEC announced an output increase lower than expected. US stocks fell lower than experts predicted, changes in federal oil policies, and six pipelines were attacked by a leftist group in Mexico. In October 2007 US light crude rose above $90 for the first time, due to a combination of tensions in eastern Turkey and the reducing strength of the US dollar.

On January 2, 2008, a single trade was made at $100, but the price did not stay above $100 until late February.

Oil broke through $110 on March 12, 2008, $125 on May 9, 2008, $130 on May 21, 2008, $135 on May 22, 2008, $140 on June 26, 2008 and $145 on July 3, 2008. On July 11, 2008, oil prices rose to a new record of $147.27 following concern over recent Iranian missile tests.

On July 14, 2008, President George W. Bush lifted the executive order removing the ban on offshore drilling that had been enacted by President George H. W. Bush in 1990 and renewed by President William J. Clinton.

This action was viewed initially as only symbolic or political since the 1982 Congressional moratorium on offshore drilling was unaffected by President Bush's action. Oil prices declined by more than $20 over the next two weeks, settling around $125 a barrel on July 24, 2008. A strong contributor to this price decline was the drop in demand for oil in the US. Miles driven there in a month were down in March–May 2008 compared to 2007, with the 4% decline in May being the largest drop in history. Oil further dropped down to its lowest price in 3 months, at around $112 a barrel, on August 11, 2008, and on September 15, oil price fell below $100 for the first time in seven months. On September 24, 2008, Speaker of the House, Nancy Pelosi allowed the 26-year moratorium to expire. On October 11, oil fell as much as $8.97, or 10.17% to $77.70 per barrel as global equities slid.

Oil traded below $70 on October 16, 2008. On December 21, 2008, oil was trading at $33.87 a barrel, less than one fourth of the peak price reached four months earlier. Prices initially continued falling in 2009, descending by mid-February to below $34, but prices then steadily rose over the next two years, hitting $113.93/barrel in April 2011.

Benchmark Pricing

After the collapse of the OPEC-administered pricing system in 1985, and a short lived experiment with netback pricing, oil-exporting countries adopted a market-linked pricing mechanism. First adopted by PEMEX in 1986, market-linked pricing received wide acceptance and by 1988 became and still is the main method for pricing crude oil in international trade. The current reference, or pricing markers, are Brent, WTI, and Dubai/Oman.

Market Listings

Oil is marketed among other products in commodities markets. See above for details. Widely traded oil futures, and related natural gas futures, include:

- Petroleum
  - Nymex Crude Future
  - Dated Brent Spot
  - WTI Cushing Spot
  - Nymex Heating Oil Future
  - Nymex RBOB Gasoline Future
- Natural gas
  - Nymex Henry Hub Future
  - Henry Hub Spot
  - New York City Gate Spot.

Most of the above oil futures have delivery dates in all 12 months of the year.

Speculation

The surge in oil prices in the past several years has led some experts to argue that at least some of the rise is due to speculation in the futures markets. This has led to an investigation, which reached an interim conclusion that speculation was largely not responsible for the rise. Economist James K. Galbraith believes that much of the rise is due to the "Enron loophole" drafted in a rider by former Texas senator Phil Gramm, which allowed energy futures to avoid Commodity Futures Trading Commission oversight.

Galbraith cites Masters, a hedge fund manager, who observed that index speculation tied to commodities by pension funds and other investment vehicles rose from $13 billion in 2003 to $250 billion in 2008. Galbraith observed that with Goldman Sachs predicting a rise in the price to $200 and Gazprom $250, suppliers may react to the rise by restricting supply until they can sell their product at a higher price. In 2009, Seismic Micro-Technology conducted a survey of geophysicists and geologists about the future of crude oil. Of the survey participants 80 percent predicted the price for a barrel of oil will rise to be somewhere between $50 and $100 per barrel by June 2010. Another 50 percent saying it will rise even further to $100 to $150 a barrel in the next five years.

Oil prices could go to $200- $300 a barrel if the world's top crude exporter Saudi Arabia is hit by serious political unrest, according to former Saudi oil minister Sheikh Yamani. Yamani has said that underlying discontent remained unresolved in Saudi Arabia. "If something happens in Saudi Arabia it will go to $200 to $300. I don't expect this for the time being, but who would have expected Tunisia?" Yamani told Reuters on the sidelines of a conference of the Centre for Global Energy Studies (CGES) which he chaired on April 5th 2011.

Futures Investigation

The U.S. Commodity Futures Trading Commission (CFTC) announced "Multiple Energy Market Initiatives" on May 29, 2008. Part 1 is "Expanded International Surveillance Information for Crude Oil Trading." The CFTC announcement stated it has joined with the United Kingdom Financial Services Authority and ICE Futures Europe in order to expand surveillance and information sharing of various futures contracts. This announcement has received wide coverage in the financial press, with speculation about oil futures price manipulation. The interim report by the Interagency Task Force, released in July, found that speculation had not caused significant changes in oil prices and that fundamental supply and demand factors provide the best explanation for the crude oil price increases. The report found that the primary reason for the price increases was that the world economy had expanded at its fastest pace in decades, resulting in substantial increases in the demand for oil, while the oil production grew sluggishly, compounded by production shortfalls in oil-exporting countries.

The report stated that as a result of the imbalance and low price elasticity, very large price increases occurred as the market attempted to balance scarce supply against growing demand, particularly in the last three years. The report forecast that this imbalance would persist in the future, leading to continued upward pressure on oil prices, and that large or rapid movements in oil prices are likely to occur even in the absence of activity by speculators. The task force continues to analyze commodity markets and intends to issue further findings later in the year.

7

Oil Reserves

The total estimated amount of oil in an oil reservoir, including both producible and non-producible oil, is called *oil in place*. However, because of reservoir characteristics and limitations in petroleum extraction technologies, only a fraction of this oil can be brought to the surface, and it is only this producible fraction that is considered to be *reserves*.

The ratio of producible oil reserves to total oil in place for a given field is often referred to as the *recovery factor*. Recovery factors vary greatly among oil fields. The recovery factor of any particular field may change over time based on operating history and in response to changes in technology and economics. The recovery factor may also rise over time if additional investment is made in enhanced oil recovery techniques such as gas injection, surfactants injection, water-flooding, or microbial enhanced oil recovery.

Based on data from OPEC at the beginning of 2011 the highest proved oil reserves including non-conventional oil deposits are in Venezuela (20 per cent of global reserves), Saudi Arabia (18 %,of global reserves), Iran (9 %).

Because the geology of the subsurface cannot be examined directly, indirect techniques must be used to estimate the size and recoverability of the resource. While new technologies have increased the accuracy of these techniques, significant uncertainties still remain. In general, most early estimates of the reserves of an oil field are conservative and tend to grow with time. This phenomenon is called *reserves growth.*

Many oil-producing nations do not reveal their reservoir engineering field data and instead provide unaudited claims for their oil reserves. The numbers disclosed by some national governments are suspected of being manipulated for political reasons.

Classifications

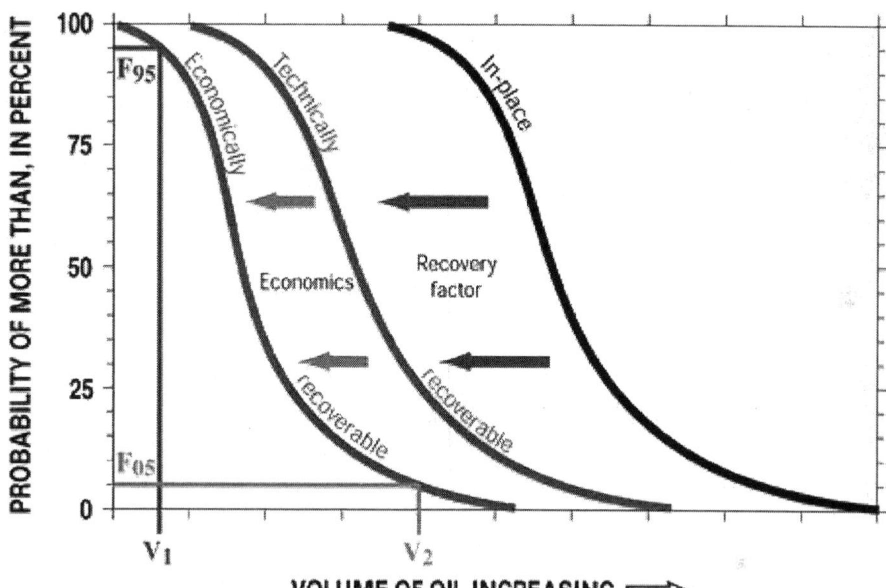

Figure: *Schematic graph illustrating petroleum volumes and probabilities. Curves represent categories of oil in assessment. There is a 95% chance (i.e., probability, F95) of at least volume V1 of economically recoverable oil, and there is a 5% chance (F05) * remaining in the ground All reserve estimates involve uncertainty, depending on the amount of reliable geologic and engineering data available and the interpretation of those data. The relative degree of uncertainty can be expressed by dividing reserves into two principal classifications—"proven" (or "proved") and "unproven" (or "unproved"). Unproven reserves can further be divided into two subcategories—"probable" and "possible"—to indicate the relative degree of uncertainty about their existence. The most commonly accepted definitions of these are based on those approved by the Society of Petroleum Engineers (SPE) and the World Petroleum Council (WPC) in 1997.*

Proven reserves are those reserves claimed to have a *reasonable certainty* (normally at least 90% confidence) of being recoverable under existing economic and political conditions, with existing technology. Industry specialists refer to this as P90 (i.e., having a 90%

certainty of being produced). Proven reserves are also known in the industry as 1P. Proven reserves are further subdivided into "proven developed" (PD) and "proven undeveloped" (PUD).

PD reserves are reserves that can be produced with existing wells and perforations, or from additional reservoirs where minimal additional investment (operating expense) is required. PUD reserves require additional capital investment (e.g., drilling new wells) to bring the oil to the surface.

Until December 2009 "1P" proven reserves were the only type the U.S. Securities and Exchange Commission allowed oil companies to report to investors. Companies listed on U.S. stock exchanges must substantiate their claims, but many governments and national oil companies do not disclose verifying data to support their claims. Since January 2010 the SEC now allows companies to also provide additional optional information declaring "2P" (both proven and probable) and "3P" (proven + probable + possible) provided the evaluation is verified by qualified third party consultants, though many companies choose to use 2P and 3P estimates only for internal purposes.

Unproven Reserves

Unproven reserves are based on geological and/or engineering data similar to that used in estimates of proven reserves, but technical, contractual, or regulatory uncertainties preclude such reserves being classified as proven. Unproven reserves may be used internally by oil companies and government agencies for future planning purposes but are not routinely compiled. They are sub-classified as *probable* and *possible.*

Probable reserves are attributed to known accumulations and claim a 50% confidence level of recovery. Industry specialists refer to them as P50 (i.e., having a 50% certainty of being produced). These reserves are also referred to in the industry as 2P (proven plus probable).

Possible reserves are attributed to known accumulations that have a less likely chance of being recovered than probable reserves. This term is often used for reserves which are claimed to have at least a 10% certainty of being produced (P10). Reasons for classifying reserves as possible include varying interpretations of geology, reserves not producible at commercial rates, uncertainty due to reserve infill (seepage from adjacent areas) and projected reserves based on future recovery methods. They are referred to in the industry as 3P (proven plus probable plus possible).

Strategic Petroleum Reserves

Many countries maintain government-controlled oil reserves for both economic and national security reasons. According to the United States Energy Information Administration, approximately 4.1 billion barrels (650,000,000 m^3) of oil are held in strategic reserves, of which 1.4 billion is government-controlled. These reserves are generally not counted when computing a nation's oil reserves.

Resources

A more sophisticated system of evaluating petroleum accumulations was adopted in 2007 by the Society of Petroleum Engineers (SPE), World Petroleum Council (WPC), American Association of Petroleum Geologists (AAPG), and Society of Petroleum Evaluation Engineers (SPEE). It incorporates the 1997 definitions for reserves, but adds categories for *contingent resources* and *prospective resources.*

Contingent resources are those quantities of petroleum estimated, as of a given date, to be potentially recoverable from *known* accumulations, but the applied project(s) are not yet considered mature enough for commercial development due to one or more contingencies.

Contingent resources may include, for example, projects for which there are currently no viable markets, or where commercial recovery is dependent on technology under development, or where evaluation of the accumulation is insufficient to clearly assess commerciality.

Prospective resources are those quantities of petroleum estimated, as of a given date, to be potentially recoverable from *undiscovered* accumulations by application of future development projects. Prospective resources have both an associated chance of discovery and a chance of development.

The United States Geological Survey uses the terms *technically* and *economically* recoverable resources when making its petroleum resource assessments. Technically recoverable resources represent that proportion of assessed in-place petroleum that may be recoverable using current recovery technology, without regard to cost. Economically recoverable resources are technically recoverable petroleum for which the costs of discovery, development, production, and transport, including a return to capital, can be recovered at a given market price.

Unconventional resources exist in petroleum accumulations that are pervasive throughout a large area. Examples include extra heavy oil, natural bitumen, and oil shale deposits. Unlike conventional resources, in which the petroleum is recovered through wellbores and typically requires minimal processing prior to sale, unconventional resources require specialized extraction technology to produce.

For example, steam and/or solvents are used to mobilize bitumen for in-situ recovery. Moreover, the extracted petroleum may require significant processing prior to sale (e.g., bitumen upgraders). The total amount of unconventional oil resources in the world considerably exceeds the amount of conventional oil reserves, but are much more difficult and expensive to develop.

Estimation Techniques

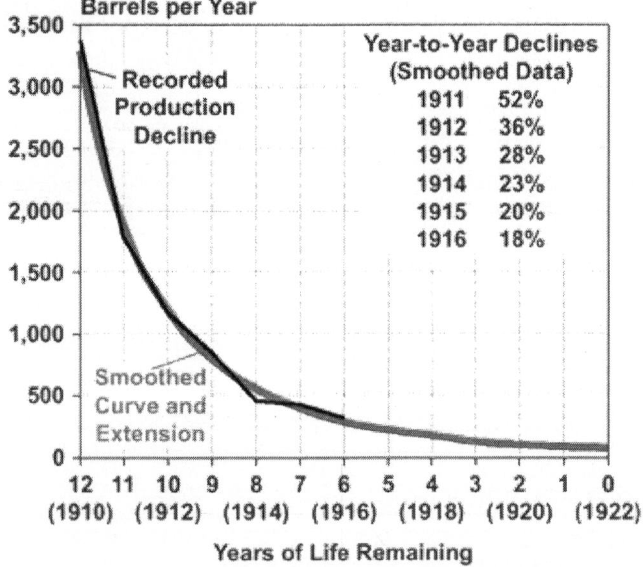

Figure: *Example of a production decline curve for an individual well*

The amount of oil in a subsurface reservoir is called *oil in place* (OIP). Only a fraction of this oil can be recovered from a reservoir. This fraction is called the *recovery factor*. The portion that can be recovered is considered to be a reserve. The portion that is not recoverable is not included unless and until methods are implemented to produce it.

There are a number of different methods of calculating oil reserves. These methods can be grouped into three general categories: volumetric, material balance, and production performance. Each method has its advantages and drawbacks.

Volumetric Method

Volumetric methods attempt to determine the amount of oil in place by using the size of the reservoir as well as the physical properties of its rocks and fluids. Then a recovery factor is assumed, using assumptions from fields with similar characteristics. OIP is multiplied by the recovery factor to arrive at a reserve number. Current recovery factors for oil fields around the world typically range between 10 and 60 percent; some are over 80 percent. The wide variance is due largely to the diversity of fluid and reservoir characteristics for different deposits. The method is most useful early in the life of the reservoir, before significant production has occurred.

Materials Balance Method

The *materials balance method* for an oil field uses an equation that relates the volume of oil, water and gas that has been produced from a reservoir and the change in reservoir pressure to calculate the remaining oil. It assumes that, as fluids from the reservoir are produced, there will be a change in the reservoir pressure that depends on the remaining volume of oil and gas.

The method requires extensive pressure-volume-temperature analysis and an accurate pressure history of the field. It requires some production to occur (typically 5% to 10% of ultimate recovery), unless reliable pressure history can be used from a field with similar rock and fluid characteristics.

Production Decline Curve Method

The *decline curve method* uses production data to fit a decline curve and estimate future oil production. The three most common forms of decline curves are exponential, hyperbolic, and harmonic. It is assumed that the production will decline on a reasonably smooth curve, and so allowances must be made for wells shut in and production restrictions.

The curve can be expressed mathematically or plotted on a graph to estimate future production. It has the advantage of (implicitly) including all reservoir characteristics. It requires a sufficient history to establish a statistically significant trend, ideally when production is not curtailed by regulatory or other artificial conditions.

Reserves Growth

Experience shows that initial estimates of the size of newly discovered oil fields are usually too low. As years pass, successive estimates of the ultimate recovery of fields tend to increase. The term *reserve growth* refers to the typical increases in estimated ultimate recovery that occur as oil fields are developed and produced.

Estimated Reserves by Country

Summary of Reserve Data as of 2011

Country	*Reserves*		*Production*		*Reserve life*[1]
	10^9 bbl	*10^9 m^3*	*10^6 bbl/d*	*10^3 m^3/d*	*years*
Saudi Arabia	264.52	42.055	8.9	1,410	81
Venezuela	211.5	33.63	2.1	330	129
Canada	175	27.8	2.7	430	178
Iran	151.2	24.04	4.1	650	101
Iraq	143.1	22.75	2.4	380	163
Kuwait	101.5	16.14	2.3	370	121
United Arab Emirates	97.8	15.55	2.4	380	112
Russia	74.2	11.80	9.7	1,540	21
Libya	47	7.5	1.7	270	76
Nigeria	37	5.9	2.5	400	41
Kazakhstan	30	4.8	1.5	240	55
Qatar	25.41	4.040	1.1	170	63
China	20.35	3.235	4.1	650	14
United States	19.12	3.040	5.5	870	10
Angola	13.5	2.15	1.9	300	19
Algeria	13.42	2.134	1.7	270	22
Brazil	13.2	2.10	2.1	330	17
Total of top seventeen reserves	1,324	210.5	56.7	9,010	64

Notes: 1 Reserve to Production ratio (in years), calculated as reserves / annual production. Although the IEA insists on Canada's Reserves as being listed as 178 billion barrels, many experts including CEO of Shell Canada, Clive Mather estimate it to actually be 2 Trillion barrels or more, essentially 8 times more than Saudi Arabia.3 Iraq is estimated to be the first oil reserve in the world with more than 360 billion barrels.4 Most of oil reserves on Venezuela are extra-heaby petrolium, with high amounts of sulfur.

OPEC Countries

There are doubts about the reliability of official OPEC reserves estimates, which are not provided with any form of audit or verification that meet external reporting standards.

Since a system of country production quotas was introduced in the 1980s, partly based on reserves levels, there have been dramatic increases in reported reserves among OPEC producers. In 1983, Kuwait increased its proven reserves from 67 Gbbl ($10.7\times10^{\triangle 9}$ m^3) to 92 Gbbl ($14.6\times10^{\triangle 9}$ m^3). In 1985–86, the UAE almost tripled its reserves from 33 Gbbl ($5.2\times10^{\triangle 9}$ m^3) to 97 Gbbl ($15.4\times10^{\triangle 9}$ m^3). Saudi Arabia raised its reported reserve number in 1988 by 50%. In 2001–02, Iran raised its proven reserves by some 30% to 130 Gbbl ($21\times10^{\triangle 9}$ m^3), which advanced it to second place in reserves and ahead of Iraq. Iran denied accusations of a political motive behind the readjustment, attributing the increase instead to a combination of new discoveries and improved recovery. No details were offered of how any of the upgrades were arrived at.

The following table illustrates these rises.

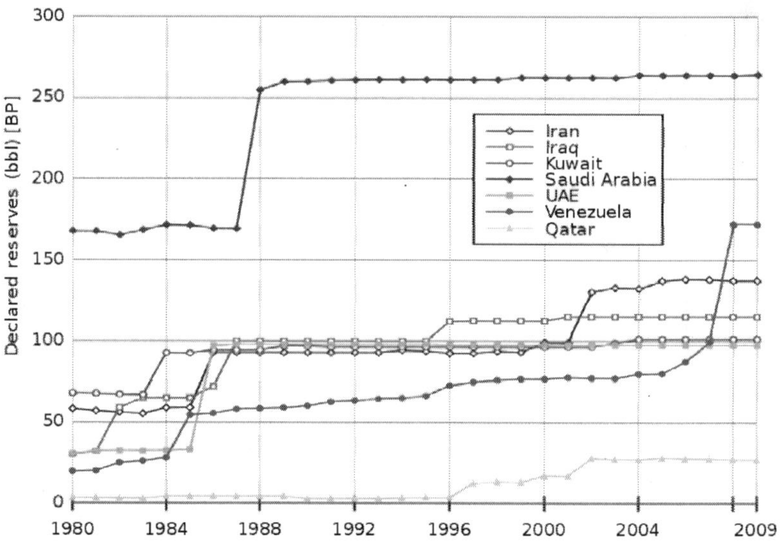

Figure: *Oil reserves of OPEC 1980–2005*

Declared reserves of major Opec Producers (billion of barrels)

BP Statistical Review - June 2009

OPEC Annual Statistical Bulletin 2010/2011

Year	*Iran*	*Iraq*	*Kuwait*	*Saudi Arabia*	*UAE*	*Venezuela*	*Libya*	*Nigeria*
1980	58.3	30.0	67.9	168.0	30.4	19.5	20.3	16.7
1981	57.0	32.0	67.7	167.9	32.2	19.9	22.6	16.5
1982	56.1	*59.0*	67.2	165.5	32.4	24.9	22.2	16.8
1983	55.3	65.0	67.0	168.8	32.3	25.9	21.8	16.6
1984	58.9	65.0	*92.7*	171.7	32.5	28.0	21.4	16.7
1985	59.0	65.0	92.5	171.5	33.0	*54.5*	21.3	16.6
1986	*92.9*	72.0	94.5	169.7	*97.2*	55.5	22.8	16.1
1987	92.9	*100.0*	94.5	169.6	98.1	58.1	22.8	16.0
1988	92.9	100.0	94.5	*255.0*	98.1	58.5	22.8	16.0
1989	92.9	100.0	97.1	260.1	98.1	59.0	22.8	16.0
1990	92.9	100.0	97.0	260.3	98.1	60.1	22.8	17.1
1991	92.9	100.0	96.5	260.9	98.1	62.6	22.8	20.0
1992	92.9	100.0	96.5	261.2	98.1	63.3	22.8	21.0
1993	92.9	100.0	96.5	261.4	98.1	64.4	22.8	21.0
1994	94.3	100.0	96.5	261.4	98.1	64.9	22.8	21.0
1995	93.7	100.0	96.5	261.5	98.1	66.3	29.5	20.8
1996	92.6	112.0	96.5	261.4	97.8	72.7	29.5	20.8
1997	92.6	112.5	96.5	261.5	97.8	74.9	29.5	20.8
1998	93.7	112.5	96.5	261.5	97.8	76.1	29.5	22.5
1999	93.1	112.5	96.5	262.8	97.8	76.8	29.5	29.0
2000	99.5	112.5	96.5	262.8	97.8	76.8	36.0	29.0
2001	99.1	115.0	96.5	262.7	97.8	77.7	36.0	31.5
2002	*130.7*	115.0	96.5	262.8	97.8	77.3	36.0	34.3
2003	133.3	115.0	99.0	262.7	97.8	77.2	39.1	35.3
2004	132.7	115.0	101.5	264.3	97.8	79.7	39.1	35.9
2005	137.5	115.0	101.5	264.2	97.8	80.0	41.5	36.2
2006	138.4	115.0	101.5	264.3	97.8	87.3	41.5	36.2
2007	138.2	115.0	101.5	264.2	97.8	99.4	43.7	36.2
2008	137.6	115.0	101.5	264.1	97.8	172.3	43.7	36.2
2009	137.0	115.0	101.5	264.6	97.8	211.1	46.4	36.2
2010	151.2	143.1	101.5	264.5	97.8	296.5	47.1	36.2

The sudden revisions in OPEC reserves, totalling nearly 300 bn barrels, have been much debated. Some of it is defended partly by the shift in ownership of reserves away from international oil companies, some of whom were obliged to report reserves under conservative US Securities and Exchange Commission rules.

The most prominent explanation of the revisions is prompted by a change in OPEC rules which set production quotas (partly) on reserves. In any event, the revisions in official data had little to do with the actual discovery of new reserves.

Total reserves in many OPEC countries hardly changed in the 1990s. Official reserves in Kuwait, for example, were unchanged at 96.5 Gbbl ($15.34\times10^{\triangle 9}$ m^3) (including its share of the Neutral Zone) from 1991 to 2002, even though the country produced more than 8 Gbbl ($1.3\times10^{\triangle 9}$ m^3) and did not make any important new discoveries during that period.

The case of Saudi Arabia is also striking, with proven reserves estimated at between 260 and 264 billion barrels (4.20×10^{10} m^3) in the past 18 years, a variation of less than 2%, while extracting approximately 60 billion barrels (9.5×10^{9} m^3) during this period.

Sadad al-Huseini, former head of exploration and production at Saudi Aramco, estimates 300 Gbbl ($48\times10^{\triangle 9}$ m^3) of the world's 1,200 Gbbl ($190\times10^{\triangle 9}$ m^3) of proven reserves should be recategorized as speculative resources, though he did not specify which countries had inflated their reserves.

Dr. Ali Samsam Bakhtiari, a former senior expert of the National Iranian Oil Company, has estimated that Iran, Iraq, Kuwait, Saudi Arabia and the United Arab Emirates have overstated reserves by a combined 320–390bn barrels and has said, "As for Iran, the usually accepted official 132 billion barrels (2.10×10^{10} m^3) is almost one hundred billion over any realistic assay." *Petroleum Intelligence Weekly* reported that official confidential Kuwaiti documents estimate reserves of Kuwait were only 48 billion barrels ($7.6\times10^{\triangle 9}$ m^3), of which half were proven and half were possible. The combined value of proven and possible is half of the official public estimate of proven reserves.

In July 2011, OPEC's Annual Statistical Review showed Venezuela's reserves to be larger than Saudi Arabia's.

Prospective Resources

Arctic Prospective Resources: A 2008 United States Geological Survey estimates that areas north of the Arctic Circle have 90 billion barrels (1.4×10^{10} m^3) of undiscovered, technically recoverable oil and 44 billion barrels (7.0×10^9 m^3) of natural gas liquids in 25 geologically defined areas thought to have potential for petroleum. This represented 13% of the expected undiscovered oil in the world. Of the estimated totals, more than half of the undiscovered oil resources were estimated to occur in just three geologic provinces—Arctic Alaska, the Amerasia Basin, and the East Greenland Rift Basins. More than 70% of the mean undiscovered oil resources was estimated to occur in five provinces: Arctic Alaska, Amerasia Basin, East Greenland Rift Basins, East Barents Basins, and West Greenland–East Canada. It was further estimated that approximately 84% of the oil and gas would occur offshore. The USGS did not consider economic factors such as the effects of permanent sea ice or oceanic water depth in its assessment of undiscovered oil and gas resources. This assessment was lower than a 2000 survey, which had included lands south of the Arctic Circle.

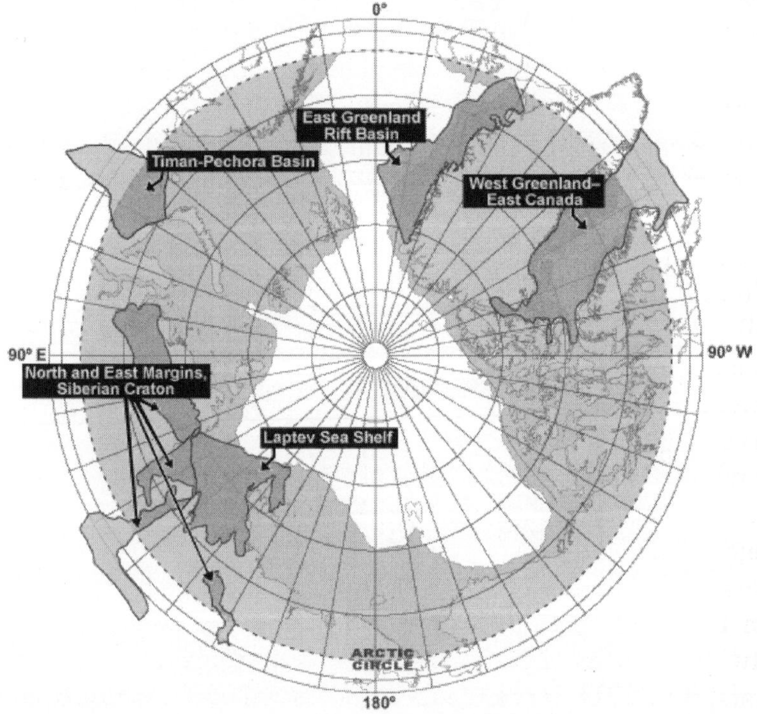

Figure: *Location of Arctic Basins assessed by the USGS*

Miscellaneous Prospective Resources

In October 2009, the USGS updated the Orinoco tar sands (Venezuela) value to 513 billion barrels (8.16×10^{10} m^3).

The Exclusive Economic Zone is a delineated offshore area, mostly to the west and north of Cuba, which, under international agreements, is owned by Cuba. This 112,000-square-kilometer zone has been divided into 59 exploration blocks. A 2004 joint partnership between a Spanish oil company and Cuba's state oil company (CUPET) estimated Cuba's offshore reserves to be able to produce ultimately between 4.6 and 9.3 billion barrels of crude oil.

The US Geological Survey (USGS) estimates that Cuba has resources up to 9 billion barrels (1.4×10^9 m^3) of oil. In October 2008, the Cuban government announced that it had discovered oil basins that would double its total oil resources to 20 billion barrels (3.2×10^9 m^3). Note that, consistent with the definitions of reserves and resources as given in the above sections, at this stage (June 2009), given no commercial discoveries to date, all recoverable oil estimates in the Exclusive Economic Zone (EEZ) are prospective resources estimates and not reserves estimates.

Global Strategic Petroleum Reserves

Global strategic petroleum reserves ("GSPR") refer to crude oil inventories (or stockpiles) held by the government of a particular country, as well as private industry, for the purpose of providing economic and national security during an energy crisis. According to the United States Energy Information Administration, approximately 4.1 billion barrels (650,000,000 m^3) of oil are held in strategic reserves, of which 1.4 billion is government-controlled. The remainder is held by private industry. At the moment the US Strategic Petroleum Reserve is one of the largest strategic reserves, with much of the remainder held by the other 26 members of the International Energy Agency. Other non-IEA countries have begun creating their own strategic petroleum reserves, with China being the largest of these new reserves. Since current consumption levels are neighbouring 0.1 billion barrels (16,000,000 m^3) per day, in the case of a dramatic worldwide drop in oil field output as suggested by some peak oil analysts, the strategic petroleum reserves are unlikely to last for more than a few months.

International Energy Agency Reserves

According to a March 2001 agreement, all 28 members of the International Energy Agency must have a strategic petroleum reserve equal to 90 days of prior year's net oil imports for their respective country. Only net-exporter members of the IEA are exempt from the reserve requirement. The exempt countries are Canada, Denmark, Norway, and the United Kingdom. However, Denmark and the UK have both recently created strategic reserves due to their requirements as European Union members.

Forward Commercial Storage Agreements

To allow oil-exporting countries increased flexibility in their production quotas, there has been an increased movement towards forward commercial storage agreements. These agreements allow petroleum to be stored at an oil-importing country, however the reserves are technically under the control of the oil-exporting country. Oil importing countries benefit from the close access to the commercial reserves, while reducing the costs of access.

Emergency Oil Sharing Agreements

In addition to maintaining a domestic stockpile of petroleum, several countries also have agreements to share their stockpiles in the event of an emergency.

The Japan, New Zealand and South Korea Agreement

In mid-2007 Japan announced a program to share its strategic reserve with other countries in its region. Negotiations are currently underway with New Zealand on an emergency oil-sharing program whereby Japan would make available for purchase its strategic reserves. In an emergency New Zealand would pay the market price plus negotiated option fees for the amount of oil previously held for them by Japan.

South Korea and Japan have also agreed to share their oil reserves in case of an emergency.

The United States and Israel Agreement

According to the 1975 Second Sinai withdrawal document signed by the United States and Israel, in an emergency the U.S. is obligated to make oil available for sale to Israel for up to 5 years.

The France, Germany, and Italy Agreement

France, Germany and Italy have an oil-sharing agreement to allow each other to purchase their strategic reserves in the event of an emergency.

Africa

South Africa has an SPR. It is managed by PetroSA and the primary facility is the Saldanha Bay oil storage facility, which is a major transit point for oil shipping. Saldanha Bay's six in-ground concrete storage tanks give the facility a storage capacity of 45,000,000 barrels (7,200,000 m^3).

Malawi is considering creating a 21-day reserve of fuel, which is an expansion from the current five day reserve. The government has begun planning for storage facilities in the provinces of Chipoka and Mchinji as well as Kamuzu International Airport.

Kenya is setting up a Strategic Fuel Reserve, similar to that of cereals. The strategic stocks would be procured by the National Oil Corporation of Kenya and stored by the Kenya Pipeline Company Limited.

Asia

China: In 2007 China announced an expansion of their crude reserves into a two part system. Chinese reserves would consist of a government-controlled strategic reserve complemented by mandated commercial reserves. The government-controlled reserves are being completed in three phases. Phase one consisted of a 101,900,000 barrels (16,200,000 m^3) reserve, mostly completed by the end of 2008. The second phase of the government-controlled reserves with an additional 170,000,000 barrels (27,000,000 m^3) will be completed by 2011. Recently, Zhang Guobao the head of the National Energy Administration also stated that there will be a third phase that will expand reserves by 204,000,000 barrels (32,400,000 m^3) with the goal of increasing China's SPR to 90 days of supply by 2020.

The planned state reserves of 475,900,000 barrels (75,660,000 m^3) plus the planned enterprise reserves of 209,440,000 barrels (33,298,000 m^3) will provide around 90 days of consumption or a total of 684,340,000 barrels (108,801,000 m^3).

India: India has begun the development of a strategic crude oil reserve sized at 37,400,000 barrels (5,950,000 m^3), enough for two weeks of consumption. Petroleum stocks have been transferred from the Indian Oil Corporation (IndianOil) to the Oil Industry Development Board (OIDB). The OIDB then created the Indian Strategic Petroleum Reserves Ltd (ISPRL) to serve as the controlling government agency for the strategic reserve.

The facilities are:

- Mangalore, State of Karnataka. Capacity of 11.22 million barrels (1,784,000 m^3).
- Padur village, Udipi in the state of Karnataka. Capacity of 18.7 million barrels (2,970,000 m^3).
- Visakhapatnam, State of Andhra Pradesh. Capacity of 7.48 million barrels (1,189,000 m^3).

Japan: As of 2010 Japan has an SPR composed of the following three types of stockpiles:

- State controlled reserves of petroleum at eleven different locations totalling 324,000,000 barrels (51,500,000 m^3). *All numbers, unless otherwise cited, come from p. 177 of this document:*
  - Tomakomai Eastern Oil Reserve Storage Base, 55 storage tanks, total capacity 34 million barrels (5,400,000 m^3).
  - Mutsu-Ogawara Storage Base, 53 storage tanks, total capacity 31 million barrels (4,900,000 m^3).
  - Kuji Storage Base, 3 storage tanks, total capacity 10.5 million barrels (1,670,000 m^3).
  - Akita Storage Base, 15 storage tanks, total capacity 23.4 million barrels (3,720,000 m^3).
  - Fukui Storage Base, 27 storage tanks, total capacity 17.9 million barrels (2,850,000 m^3).
  - Kikuma Underground Petroleum Storage Facility, 8 storage tanks, total capacity 8.9 million barrels (1,410,000 m^3).
  - Shirashima Storage Facility, 8 tankers (4,400,000 barrels (700,000 m^3) each), total capacity 35.2 million barrels (5,600,000 m^3).
  - Kamigotou Storage Base, 7 storage tanks, total capacity 21.45 million barrels (3,410,000 m^3).
  - Kushikino Storage Base, 3 storage tanks, total capacity 10.5 million barrels (1,670,000 m^3).

  - o Shibushi Storage Base, 40 storage tanks, total capacity 27.6 million barrels (4,390,000 m^3).
  - o Kagoshima former Nippon Oil facility, 4,000,000 barrels (640,000 m^3). This is a forward commercial storage facility with Abu Dhabi.
- Privately held reserves of petroleum held "in accordance with the Petroleum Stockpiling Law" of 129,000,000 barrels (20,500,000 m^3).
- Privately held reserves of petroleum products for another 130,000,000 barrels (21,000,000 m^3).

The state stockpile and the privately held stockpiles total about 583,000,000 barrels (92,700,000 m^3). The Japanese SPR is run by the Japan Oil, Gas and Metals National Corporation.

South Korea: As part of the government's energy security efforts, South Korea holds strategic oil reserves to protect against oil supply disruptions. The country's strategic oil reserve program is managed by the Korea National Oil Corporation, which reports that its system has the capacity to store 116,000,000 barrels (18,400,000 m^3) of oil. As of April 2007, KNOC held 76,000,000 barrels (12,100,000 m^3) of oil in its strategic stockpiles, 64,000,000 barrels (10,200,000 m^3) of crude oil and 12,000,000 barrels (1,900,000 m^3) of petroleum products. This total amounts to approximately 34 days of net import cover, according to 2006 estimates of demand. KNOC has plans to expand the country's strategic storage capacity from 116,000,000 barrels (18,400,000 m^3) to 146,000,000 barrels (23,200,000 m^3) by 2009, and to fill the emergency reserves to 141,000,000 barrels (22,400,000 m^3) by 2010.

Others: The Philippines has begun plans for a National Petroleum Strategic Reserve by 2010 with an approximate size of 30,000,000 barrels (4,800,000 m^3).

Russia has begun plans for a strategic petroleum reserve. Analysts estimate the size of the Russian SPR would be around 78,000,000 barrels (12,400,000 m^3).

Singapore has an SPR composed of 31.8 million barrels (5,060,000 m^3) of crude oil with an additional 64.5 million barrels (10,250,000 m^3) of oil products for a total of 96,300,000 barrels (15,310,000 m^3).

Taiwan has an SPR with a 1999 reported size of 13,000,000 barrels (2,100,000 m^3). Taiwan's refiners (Kaohsiung 270,000 bbl/d

(43,000 m³/d); Ta-Lin 300,000 bbl/d (48,000 m³/d); Tao-Yuan 200,000 bbl/d (32,000 m³/d); Mailiao 150,000 bbl/d) are also required to store at least 30 days of petroleum stocks. As of 2005, these mandated commercial reserves total 27,600,000 barrels (4,390,000 m³) of strategic petroleum stocks.

Thailand has increased the size of its SPR from 60 days to 70 days of consumption in 2006.

Pakistan has begun plans for a 20 day emergency reserve.

Europe

European Union: In the European Union, according to Council Directive 68/414/EEC of 20 December 1968, all 27 members must have a strategic petroleum reserve within the territory of the E.U. equal to at least 90 days average daily internal consumption.

The Czech Republic has a four tank SPR facility in Nelahozeves run by the company CR Mero. The Czech SPR is equal to 100 days of consumption or 20,300,000 barrels (3,230,000 m³).

Denmark has a reserve of 81 days of consumption, equal to about 1,4 million tonnes of oil products. Not counting reserves held by the military defence.

Finland has an SPR with an approximate size of 62,400,000 barrels (9,920,000 m³). France has an SPR with an approximate size of 65,000,000 barrels (10,300,000 m³). As of 2000 jet fuel stocks were required for at least 55 days of consumption, with half of those stocks controlled by the *Société Anonyme de Gestion des Stocks de Sécurité* (SAGESS) and the other half controlled by producers.

Germany created the Federal Oil Reserve in 1970, stored in the Etzel salt caverns near Wilhelmshaven in northern Germany, with an initial size of 70 million barrels (11,000,000 m³). The current German Federal Oil Reserve and the Erdolbevorratungsverband (EBV) (the German stockholding company) mandates that refiners must keep 90 days of stock on hand, giving Germany an approximate reserve size of 250,000,000 barrels (40,000,000 m³) as of 1997. The German SPR is the largest in Europe.

Hungary has an SPR with approximately 90 days of consumption or 11,880,000 barrels (1,889,000 m³).

Ireland has approximately 31 days of oil stocks in Ireland and another 9 days of oil stocks held in fellow EU members states.

Additionally, they have stock tickets (contracts with a 3rd party where the government has the option to purchase in the event of an emergency) and stocks held by large industry or large consumers. On average Ireland has approximately 100 days of oil available.

Poland has an SPR with approximately 70 days of consumption. Another facility holding 20 additional days of consumption is scheduled to be completed in 2008. Poland also requires oil companies to maintain reserves sufficient for 73 days of production.

Portugal has an SPR with an approximate size of 22,440,000 barrels (3,568,000 m^3). Slovakia has an SPR with an approximate size of 748,000 barrels (118,900 m^3). Spain has an SPR with an approximate size of 120,000,000 barrels (19,000,000 m^3). Sweden has an SPR with an approximate size of 13,290,000 barrels (2,113,000 m^3).

The United Kingdom has created a strategic reserve, the size is unknown.

Russia: Russia has begun accumulating strategic reserves of refined products which will be held by Rosneftegaz, the state-owned company. The reserves will be held at commercial refineries, Transneft facilities, and state reserve facilities. The current planned size is 14,665,982 barrels (2,331,704.8 m^3).

Switzerland: Switzerland has SPRs consisting of gas, diesel, jet fuel and heating oil for 4.5 months of consumption. The reserves were created in the 1940s and were used for the first time in 2005 following Hurricane Katrina.

Middle East

Iran: In April 2006 the Fars News Agency reported that Iran has begun plans to create an SPR. The National Iranian Oil Company (NIOC) has begun construction of 15 crude oil storage tanks with a planned capacity of 10,000,000 barrels (1,600,000 m^3). In August 2008, Iran announced plans to expand their SPR with a new facility on Kharg Island containing 4 tanks holding 1,000,000 barrels (160,000 m^3) each. Iran's SPR facilities are:

- Ahwaz. 4 storage tanks, total capacity 2 million barrels (320,000 m^3).
- Omidiyeh. 3 storage tanks, total capacity 3 million barrels (480,000 m^3).
- Goureh. 6 storage tanks, total capacity 4 million barrels (640,000 m^3).

- Sirri Island. 1 storage tank, total capacity 500,000 barrels (79,000 m^3).
- Bahregansar. 1 storage tank, total capacity 500,000 barrels (79,000 m^3).
- Kharg Island. 4 storage tanks, total capacity 4 million barrels (640,000 m^3). Planned facility, not operational yet.

Kuwait: Kuwait has a joint stockpile held in South Korea. The deal gives South Korea first rights to purchase the oil. The current size of the stockpile is 2 million barrels (320,000 m^3).

Others: As of 1975 Israel is believed to have a strategic oil reserve equal to 270 days of consumption.

Jordan has strategic oil reserves equal to 60 days of consumption or 6,240,000 barrels (992,000 m^3).

North America

United States: The United States has the largest reported Strategic Petroleum Reserve with a total capacity of 727,000,000 barrels (115,600,000 m^3). If completely filled, the US SPR could theoretically replace about 60 days of oil imports as the US is estimated to import approximately 12,000,000 barrels per day (1,900,000 m^3/d) of crude oil. According to the US Department of Energy the facilities maximum flow rate is limited to approximately 4,400,000 barrels per day (700,000 m^3/d) when filled to maximum, with flow rate declining as the reserve is drawn down. The U.S. facilities are in salt caverns with locations in:

- Bryan Mound - located near Freeport, Texas. Capacity of 226,000,000 barrels (35,900,000 m^3).
- Big Hill - located near Winnie, Texas. Capacity of 160,000,000 barrels (25,000,000 m^3).
- West Hackberry - located near Lake Charles, Louisiana. Capacity of 219,000,000 barrels (34,800,000 m^3).
- Bayou Choctaw - located near Baton Rouge, Louisiana. Capacity of 72,000,000 barrels (11,400,000 m^3).
- Richton, Mississippi. A planned facility.

The US has also organized the 2-million-barrel (320,000 m^3) Northeast Home Heating Oil Reserve to supply northeast homeowners during shortages.

California: The state of California is considering the creation of Strategic Fuels Reserve.

Oceania: New Zealand has a strategic reserve with a 2008 size of 170,000 tons or 1,200,000 barrels (190,000 m^3). Much of this reserve is based on ticketed option contracts with Australia, Japan, the United Kingdom and the Netherlands, which allow for guaranteed purchases of petroleum in the event of an emergency.

Strategic Petroleum Reserve

The Strategic Petroleum Reserve (SPR) is an emergency fuel store of oil maintained by the United States Department of Energy.

United States

The US SPR is the largest emergency supply in the world with the current capacity to hold up to 727 million barrels (115,600,000 m^3).

The current inventory is displayed on the SPR's website. As of May 31, 2011, the current inventory was 726.5 million barrels (115,500,000 m^3). This equates to 34 days of oil at current daily US consumption levels of 21 million barrels per day (3,300,000 m^3/d). At recent market prices ($65 a barrel as of October 2008) the SPR holds over $34.3 billion in sweet crude and approximately $51.2 billion in sour crude (assuming a $15/barrel discount for sulfur content). The total value of the crude in the SPR is approximately $85.5 billion USD. The price paid for the oil is $20.1 billion (an average of $28.42 per barrel).

Purchases of crude oil resumed in January 2009 using revenues available from the 2005 Hurricane Katrina emergency sale. The DOE purchased 10,700,000 barrels (1,700,000 m^3) at a cost of $553 million.

The United States started the petroleum reserve in 1975 after oil supplies were cut off during the 1973-74 oil embargo, to mitigate future temporary supply disruptions. According to the World Factbook, the United States imports a net 12 million barrels (1,900,000 m^3) of oil a day (MMbd), so the SPR holds about a 58-day supply. However, the maximum total withdrawal capability from the SPR is only 4.4 million barrels (700,000 m^3) per day, making it a 160 + day supply.

Facilities

The SPR management office is located in New Orleans, Louisiana.

The reserve is stored at four sites on the Gulf of Mexico, each located near a major center of petrochemical refining and processing.

Each site contains a number of artificial caverns created in salt domes below the surface.

Individual caverns within a site can be up to 1000 m below the surface, average dimensions are 60 m wide and 600 m deep, and capacity ranges from 6 to 37 million barrels (950,000 to 5,900,000 m^3). Almost $4 billion was spent on the facilities. The decision to store in caverns was made in order to reduce costs; the Department of Energy claims it is roughly 10 times cheaper to store oil below surface with the added advantages of no leaks and a constant natural churn of the oil due to a temperature gradient in the caverns. The caverns were created by drilling down and then dissolving the salt with water.

Existing:

- Bryan Mound - Freeport, Texas. 20 caverns with a storage capacity of 254 million barrels (40,400,000 m^3) with a drawdown capacity of 1.5 million barrels (240,000 m^3) per day.
- Big Hill - Winnie, Texas. Has a capacity of 160 million barrels (25,000,000 m^3) with a drawdown capacity of 1.1 million barrels (170,000 m^3) per day. This facility is planned to be expanded by 250 million barrels (40,000,000 m^3) with a new drawdown capacity of 1.5 million barrels (240,000 m^3) per day.
- West Hackberry - Lake Charles, Louisiana. Has a capacity of 227 million barrels (36,100,000 m^3) with a drawdown capacity of 1.3 million barrels (210,000 m^3) per day.
- Bayou Choctaw - Baton Rouge, Louisiana. Has a capacity of 76 million barrels (12,100,000 m^3) with a maximum drawdown rate of 550,000 barrels (87,000 m^3). This facility is planned to be expand to 109 million barrels (17,300,000 m^3) with a new drawdown capacity of 600,000 barrels (95,000 m^3) per day.

Future:

- Richton, Mississippi. This facility, if built as planned, will have a capacity of 160 million barrels (25,000,000 m^3) with a drawdown capacity of 1 million barrels (160,000 m^3) per day. The Secretary of the Energy Department, Samuel Bodman, announced the creation of this site in February 2007. This new site is currently facing some opposition.

Retired:

- Weeks Island - Iberia Parish, Louisiana (Decommissioned 1999) Capacity of 72 million barrels (11,400,000 m^3). This facility was a conventional room and pillar near-surface salt mine, formerly owned by Morton Salt. In 1993, a sinkhole formed on the site, allowing fresh water to intrude into the mine. Because of the mine's construction in salt deposits, fresh water would erode the ceiling, potentially causing the structure to fail. The mine was backfilled with salt-saturated brine. This process, which allowed for recovery of 98% of the petroleum stored in the facility, reduced the risk of further freshwater intrusion, and helped prevent the remaining oil from leaking into the aquifer that is located over the salt dome.

History

Background: Access to the reserve is determined by the conditions written into the 1975 Energy Policy and Conservation Act (EPCA), primarily to counter a severe supply interruption. The maximum removal rate, by physical constraints, is 4.4 million barrels per day (700,000 m^3/d). Oil could begin entering the marketplace 13 days after a Presidential order.

The Dept. of Energy says that it has about 59 days of import protection in the SPR. This, combined with private sector inventory protection, is estimated to equal 115 days of imports. The SPR was created following the 1973 energy crisis. The EPCA of December 22, 1975, made it policy for the U.S. to establish a reserve up to one billion barrels (159 million m^3) of petroleum.

A number of existing storage sites were acquired in 1977. Construction of the first surface facilities began in June 1977. On July 21, 1977, the first oil—approximately 412,000 barrels (65,500 m^3) of Saudi Arabian light crude—was delivered to the SPR. Fill was suspended in FY 1995 to devote budget resources to refurbishing the SPR equipment and extending the life of the complex. The current SPR sites are expected to be usable until around 2025. Fill was resumed in 1999.

Filling and Suspending the SPR

On November 13, 2001, President George W. Bush announced that the SPR would be filled, saying, "The Strategic Petroleum Reserve is an important element of our Nation's energy security. To maximize long-term protection against oil supply disruptions, I am directing the Secretary of Energy to fill the SPR up to its 700 million barrel [111,000,000 m^3] capacity."

The highest prior level was reached in 1994 with 592 million barrels (94,100,000 m^3). At the time of President Bush's directive, the SPR contained about 545 million barrels (86,600,000 m^3). Since the directive in 2001, the capacity of the SPR increased by 27 million barrels (4,300,000 m^3) due to natural enlargement of the salt caverns in which the reserves are stored. The Energy Policy Act of 2005 has since directed the Secretary of Energy to fill the SPR to the full 1-billion-barrel (160,000,000 m^3) authorized capacity, a process which will require a physical expansion of the Reserve's facilities.

On August 17, 2005, the SPR reached its goal of 700 million barrels (110,000,000 m^3), or about 96% of its now-increased 727-million-barrel (115,600,000 m^3) capacity. Approximately 60% of the crude oil in the reserve is the less desirable sour (high sulfur content) variety.

The oil delivered to the reserve is "royalty-in-kind" oil—royalties owed to the U.S. government by operators who acquire leases on the federally owned Outer Continental Shelf in the Gulf of Mexico. These royalties were previously collected as cash, but in 1998 the government began testing the effectiveness of collecting royalties "in kind" - or in other words, acquiring the crude oil itself. This mechanism was adopted when refilling the SPR began, and once filling is completed, revenues from the sale of future royalties will be paid into the federal treasury.

On April 25, 2006, President Bush announced a temporary halt to petroleum deposits to the SPR as part of a four point program to alleviate high fuel prices. On January 23, 2007, President Bush suggested in his State of the Union speech that Congress should approve expansion of the current reserve capacity to twice its current level.

In April 2008, Speaker Pelosi called on President Bush to suspend purchases of oil for the Strategic Petroleum Reserve (SPR) temporarily.

On May 12, 2008, Rep. Peter Welch (D, Vermont) and 63 co-sponsors introduced the Strategic Petroleum Reserve Fill Suspension and Consumer Protection Act bill (H.R.6022), to suspend the acquisition of petroleum for the Strategic Petroleum Reserve.

On May 16, 2008, the U.S. Department of Energy said it would halt all deliveries to the Strategic Petroleum Reserve sometime in July. This announcement came days after Congress voted to direct the Bush administration to do the same. The U.S. Department of Energy did not state when the shipments would resume.

On May 19, 2008, President Bush signed the Act passed by the Congress, which he previously opposed. On January 2, 2009, the U.S. Energy Department said that it would begin buying approximately 12,000,000 barrels (1,900,000 m^3) of crude oil to fill the Strategic Petroleum Reserve, replenishing supplies that were sold after hurricanes Katrina and Rita in 2005. The purchase will be funded by the roughly $600 million received in 2005 from the emergency sales.

Emergency sales to Israel

According to the 1975 Second Sinai withdrawal document signed by the United States and Israel, in an emergency the U.S. is obligated to make oil available for sale to Israel for up to 5 years.

Limitations

The Strategic Petroleum Reserve is exclusively a crude petroleum reserve, not a stockpile of refined petroleum fuels, such as gasoline, diesel and kerosene. Although there are small-scale (2,000,000 barrels) heating oil reserves in Connecticut, Rhode Island and New Jersey under the aegis of the Department of Energy (DOE), the federal government maintains no gasoline reserves on anything like the scale of the SPR. Consequently, while the US enjoys some protection from disruptions in oil supplies, it would have to depend on other stockpiling members of the International Energy Agency for relief from any major disruption to refinery operations. Since no new refineries have been constructed in the US for thirty years, there is little excess refining capacity. This was illustrated during Hurricane Katrina, when many of the Gulf coast oil refining complexes were disrupted for some time.

There have been suggestions that the DOE should stockpile both gasoline and jet fuel, to rectify this weakness. Some countries and zones, such as Australia, have a strategic reserve of both petroleum and petroleum products. In some cases, this includes a strategic reserve of jet fuel.

The former Secretary of Energy, Samuel Bodman, has said the Department will consider refined products as part of the expansion of between 1 billion and 1.5 billion barrels (240,000,000 m^3).

SPR Drawdowns

Petroleum Sales:

- 1985 - Test sale - 1.1 million barrels (170,000 m^3)
- 1990/91 - Desert Storm sale - 21 million barrels (3,300,000 m^3)
    - o 4 million in August 1990 test sale

- o 17 million in January 1991 Presidentially-ordered drawdown

- 1996-97 total non-emergency sales for deficit reduction - 28 million barrels (4,500,000 m^3)
- 2005 - Hurricane Katrina sale - 11 million barrels (1,700,000 m^3) Katrina shut down 95% of crude production and 88% of natural gas output in the Gulf of Mexico. This amounted to a quarter of total U.S. output. About 735 oil and natural gas rigs and platforms had been evacuated due to the hurricane.
- 2011 - Arab Spring sale - 30 million barrels (4,800,000 m^3) to offset disruptions caused by political upheaval in Libya and elsewhere in the Middle East. The amount was matched by IEA countries, for a total of 60 million barrels (9,500,000 m^3) released from stockpiles around the world.

Petroleum Exchanges and Loans

- April–May 1996 - 900,000 barrels (140,000 m^3) lent to ARCO to alleviate pipeline blockage.
- August 1998 - 11 million barrels (1,700,000 m^3) lent to PEMEX in return for 8.5 million barrels (1,350,000 m^3) of higher quality crude.
- June 2000 - 1 million barrels (160,000 m^3) lent to Citgo and Conoco in response to shipping channel blockage.
- July–August 2000 - 2.8 million barrels (450,000 m^3) to supply the Northeast Home Heating Oil Reserve.
- September–October 2000 - 30 million barrels (4,800,000 m^3) in response to a concern over low distillate levels in the Northeastern U.S.
- October 2002 - 296,000 barrels (47,100 m^3) lent to Shell Pipeline Company in advance of Hurricane Lili.
- September–October 2004 - 5.4 million barrels (860,000 m^3) lent to Astra Oil, ConocoPhillips, Placid Refining, Shell Oil Company, and Premcor after Hurricane Ivan.
- September–October 2005 - 9.8 million barrels (1,560,000 m^3) lent to ExxonMobil, Placid Refining, Valero, BP, Marathon Oil, and Total S.A. after Hurricane Katrina.
- January–February 2006 - 767,000 barrels (121,900 m^3) lent to Total Petrochemicals USA due to closure of the Sabine Neches

ship channel to deep-draft vessels after a barge accident in the channel.

- June 2006 - 750,000 barrels (119,000 m^3) of sour crude lent to ConocoPhillips and Citgo due to the closure for several days of the Calcasieu Ship Channel caused by the release of a mixture of storm water and oil. Repaid in early October 2006.
- September 2008 - 250,000 barrels (40,000 m^3) loaned to Citgo because it could not secure crude oil in the aftermath of Hurricane Gustav.
- September 2008 - 130,000 barrels (21,000 m^3) loaned to Placid Refining's Port Allen refinery and 250,000 barrels (40,000 m^3) loaned to Marathon Oil due to disruptions from Hurricane Gustav.

Northeast Home Heating Oil Reserve

The Northeast Home Heating Oil Reserve was created in July 2000 to provide a reserve of heating oil for the approximately 5.3 million households in the Northeast region of the United States that use heating oil for their homes.

History

On July 10, 2000, President of the United States Bill Clinton directed Energy Secretary Bill Richardson to establish a 2-million-barrel home heating oil component of the Strategic Petroleum Reserve in the Northeast. The intent was to create a buffer large enough to allow commercial companies to compensate for interruptions in supply or severe winter weather, but not so large as to dissuade suppliers from responding to increasing prices as a sign that more supply is needed.

8

Oil Shale

Oil shale, an organic-rich fine-grained sedimentary rock, contains significant amounts of kerogen (a solid mixture of organic chemical compounds) from which liquid hydrocarbons called shale oil (not to be confused with tight oil—crude oil occurring naturally in shales) can be produced. Shale oil is a substitute for conventional crude oil; however, extracting shale oil from oil shale is more costly than the production of conventional crude oil both financially and in terms of its environmental impact.

Deposits of oil shale occur around the world, including major deposits in the United States of America. Estimates of global deposits range from 2.8 to 3.3 trillion barrels (450×10^9 to 520×10^9 m^3) of recoverable oil. Heating oil shale to a sufficiently high temperature causes the chemical process of pyrolysis to yield a vapor. Upon cooling the vapor, the liquid shale oil—an unconventional oil—is separated from combustible oil-shale gas (the term *shale gas* can also refer to gas occurring naturally in shales). Oil shale can also be burnt directly in furnaces as a low-grade fuel for power generation and district heating or used as a raw material in chemical and construction-materials processing.

Oil shale gains attention as a potential abundant domestic source of oil whenever the price of crude oil rises. At the same time, oil-shale mining and processing raise a number of environmental concerns, such as land use, waste disposal, water use, waste-water management, greenhouse-gas emissions and air pollution. Estonia and China have well-established oil shale industries, and Brazil, Germany, Israel and Russia also utilize oil shale.

Oil shales differ from oil-*bearing* shales, shale deposits which contain petroleum (tight oil) that is sometimes produced from drilled wells. Examples of oil-*bearing* shales are the Bakken Formation, Pierre Shale, Niobrara Formation, and Eagle Ford Formation.

Geology

Oil shale, an organic-rich sedimentary rock, belongs to the group of sapropel fuels. It does not have a definite geological definition nor a specific chemical formula, and its seams do not always have discrete boundaries. Oil shales vary considerably in their mineral content, chemical composition, age, type of kerogen, and depositional history and not all oil shales would necessarily be classified as shales in the strict sense. Oil shale differs from bitumen-impregnated rocks (oil sands and petroleum reservoir rocks), humic coals and carbonaceous shale. While oil sands originate from the biodegradation of oil, heat and pressure have not (yet) transformed the kerogen in oil shale into petroleum.

Oil shale contains a lower percentage of organic matter than coal. In commercial grades of oil shale the ratio of organic matter to mineral matter lies approximately between 0.75:5 and 1.5:5. At the same time, the organic matter in oil shale has an atomic ratio of hydrogen to carbon (H/C) approximately 1.2 to 1.8 times lower than for crude oil and about 1.5 to 3 times higher than for coals. The organic components of oil shale derive from a variety of organisms, such as the remains of algae, spores, pollen, plant cuticles and corky fragments of herbaceous and woody plants, and cellular debris from other aquatic and land plants. Some deposits contain significant fossils; Germany's Messel Pit has the status of a Unesco World Heritage Site. The mineral matter in oil shale includes various fine-grained silicates and carbonates.

Geologists can classify oil shales on the basis of their composition as carbonate-rich shales, siliceous shales, or cannel shales. Another classification, known as the van Krevelen diagram, assigns kerogen types, depending on the hydrogen, carbon, and oxygen content of oil shales' original organic matter. The most commonly used classification of oil shales, developed between 1987 and 1991 by Adrian C. Hutton of the University of Wollongong, adapts petrographic terms from coal terminology. This classification designates oil shales as terrestrial, lacustrine (lake-bottom-deposited), or marine (ocean bottom-deposited), based on the environment of the initial biomass deposit. Hutton's classification scheme has proven useful in estimating the yield and composition of the extracted oil. As with all oil and gas resources, analysts distinguish between oil shale resources and oil shale reserves. "Resources" refers to all oil shale deposits, while "reserves", represents those deposits from which producers can extract oil shale economically using existing technology. Since extraction technologies develop continuously, planners can only estimate the amount of recoverable kerogen. Although resources of oil shale occur in many countries, only 33 countries possess known deposits of possible economic value. Well-explored deposits, potentially classifiable as reserves, include the Green River deposits in the western United States, the Tertiary deposits in Queensland, Australia, deposits in Sweden and Estonia, the El-Lajjun deposit in Jordan, and deposits in France, Germany, Brazil, China, southern Mongolia and Russia. These deposits have given rise to expectations of yielding at least 40 liters of shale oil per tonne of oil shale, using the Fischer Assay.

A 2005 estimate set the total world resources of oil shale at 411 gigatons — enough to yield 2.8 to 3.3 trillion barrels (450×10^9 to 520×10^9 m^3) of shale oil. This exceeds the world's proven conventional oil reserves, estimated at 1.317 trillion barrels (209.4×10^9 m^3), as of 1 January 2007. The largest deposits in the world occur in the United States in the Green River Formation, which covers portions of Colorado, Utah, and Wyoming; about 70% of this resource lies on land owned or managed by the United States federal government. Deposits in the United States constitute 62% of world resources; together, the United States, Russia and Brazil account for 86% of the world's resources in terms of shale-oil content. These figures remain tentative, with exploration or analysis of several deposits still outstanding. Professor Alan R. Carroll of University of Wisconsin–Madison regards the Upper Permian lacustrine oil-shale deposits of northwest China, absent from previous global oil shale assessments, as comparable in size to the Green River Formation.

History

Humans have used oil shale as a fuel since prehistoric times, since it generally burns without any processing. Britons of the Iron Age also used to polish it and form it into ornaments. Modern industrial mining of oil shale began in 1837 in Autun, France, followed by exploitation in Scotland, Germany, and several other countries. Operations during the 19th century focused on the production of kerosene, lamp oil, and paraffin; these products helped supply the growing demand for lighting that arose during the Industrial Revolution. Fuel oil, lubricating oil and grease, and ammonium sulfate were also produced. The European oil-shale industry expanded immediately before World War I due to limited access to conventional petroleum resources and to the mass production of automobiles and trucks, which accompanied an increase in gasoline consumption. Although the Estonian and Chinese oil-shale industries continued to grow after World War II, most other countries abandoned their projects due to high processing costs and the availability of cheaper petroleum. Following the 1973 oil crisis, world production of oil shale reached a peak of 46 million tonnes in 1980 before falling to about 16 million tonnes in 2000, due to competition from cheap conventional petroleum in the 1980s. On 2 May 1982, known in some circles as "Black Sunday", Exxon canceled its US$5 billion Colony Shale Oil Project near Parachute, Colorado because of low oil-prices and increased expenses, laying off more than 2,000 workers and leaving a trail of home-foreclosures and small-business bankruptcies. In 1986, President Ronald Reagan signed into law the Consolidated Omnibus Budget Reconciliation Act of 1985 which among other things abolished the United States' Synthetic Liquid Fuels Program.

The global oil-shale industry began to revive at the beginning of the 21st century. In 2003, an oil-shale development program restarted in the United States. Authorities introduced a commercial leasing program permitting the extraction of oil shale and oil sands on federal lands in 2005, in accordance with the Energy Policy Act of 2005.

Industry

As of 2008, industry uses oil shale in Brazil, China, Estonia and to some extent in Germany, Israel, and Russia. Several additional countries started assessing their reserves or had built experimental production plants, while others had phased out their oil shale industry. Oil shale serves for oil production in Estonia, Brazil, and China; for power generation in Estonia, China, Israel, and Germany; for cement production in Estonia, Germany, and China; and for use in chemical industries in China, Estonia, and Russia. As of 2009, 80% of oil shale used globally is extracted in Estonia.

Romania and Russia have in the past run power plants fired by oil shale, but have shut them down or switched to other fuel sources such as natural gas. Jordan and Egypt plan to construct power plants fired by oil shale, while Canada and Turkey plan to burn oil shale along with coal for power generation. Oil shale serves as the main fuel for power generation only in Estonia, where the oil-shale-fired Narva Power Plants accounted for 95% of electrical generation in 2005.

Extraction and Processing

Most exploitation of oil shale involves mining followed by shipping elsewhere, after which one can burn the shale directly to generate electricity, or undertake further processing. The most common methods of surface mining involve open pit mining and strip mining. These procedures remove most of the overlying material to expose the deposits of oil shale, and become practical when the deposits occur near the surface. Underground mining of oil shale, which removes less of the overlying material, employs the room-and-pillar method.

The extraction of the useful components of oil shale usually takes place above ground (*ex-situ* processing), although several newer technologies perform this underground (on-site or *in-situ* processing). In either case, the chemical process of pyrolysis converts the kerogen in the oil shale to shale oil (synthetic crude oil) and oil shale gas. Most conversion technologies involve heating shale in the absence of oxygen to a temperature at which kerogen decomposes (pyrolyses) into gas, condensable oil, and a solid residue. This usually takes place between 450 °C (842 °F) and 500 °C (932 °F). The process of decomposition begins at relatively low temperatures (300 °C/570 °F), but proceeds more rapidly and more completely at higher temperatures.

In-situ processing involves heating the oil shale underground. Such technologies can potentially extract more oil from a given area of land than *ex-situ* processes, since they can access the material at greater depths than surface mines can.

Several companies have patented methods for *in-situ* retorting. However, most of these methods remain in the experimental phase. One can distinguish *true in-situ* processes (TIS) and *modified in-situ* processes (MIS). *True in-situ* processes do not involve mining the oil shale. *Modified in-situ* processes involve removing part of the oil shale and bringing it to the surface for modified *in-situ* retorting in order to create permeability for gas flow in a rubble chimney. Explosives rubblize the oil-shale deposit.

Hundreds of patents for oil shale retorting technologies exist; however, only a few dozen have undergone testing. As of 2006, only four technologies remained in commercial use: Kiviter, Galoter, Fushun, and Petrosix.

Applications and Products

Industry can use oil shale as a fuel for thermal power-plants, burning it (like coal) to drive steam turbines; some of these plants employ the resulting heat for district heating of homes and businesses. Sizable oil-shale-fired power plants occur in Estonia, which has an installed capacity of 2,967 megawatts (MW), Israel (12.5 MW), China (12 MW), and Germany (9.9 MW).

In addition to its use as a fuel, oil shale may also serve in the production of specialty carbon fibers, adsorbent carbons, carbon black, phenols, resins, glues, tanning agents, mastic, road bitumen, cement, bricks, construction and decorative blocks, soil-additives, fertilizers, rock-wool insulation, glass, and pharmaceutical products. However, oil shale use for production of these items remains small or only in its experimental stages. Some oil shales yield sulfur, ammonia, alumina, soda ash, uranium, and nahcolite as shale-oil extraction byproducts. Between 1946 and 1952, a marine type of *Dictyonema* shale served for uranium production in Sillamäe, Estonia, and between 1950 and 1989 Sweden used alum shale for the same purposes. Oil shale gas has served as a substitute for natural gas, but as of 2009, producing oil shale gas as a natural-gas substitute remained economically infeasible.

The shale oil derived from oil shale does not directly substitute for crude oil in all applications. It may contain higher concentrations of olefins, oxygen, and nitrogen than conventional crude oil. Some shale oils may have higher sulfur or arsenic content. By comparison with West Texas Intermediate, the benchmark standard for crude oil in the futures-contract market, the Green River shale oil sulfur content ranges from near 0% to 4.9% (in average 0.76%), where West Texas Intermediate's sulfur content has a maximum of 0.42%. The sulfur content in shale oil from Jordan's oil shales may rise even up to 9.5%. The arsenic content, for example, becomes an issue for Green River formation oil shale. The higher concentrations of these materials means that the oil must undergo considerable upgrading (hydrotreating) before serving as oil-refinery feedstock. Above-ground retorting processes tended to yield a lower API gravity shale oil than the *in situ* processes. Shale oil serves best for producing middle-distillates such as kerosene, jet fuel, and diesel fuel. Worldwide demand for these middle distillates, particularly for diesel fuels, increased rapidly in the 1990s and 2000s. However, appropriate refining processes equivalent to hydrocracking can transform shale oil into a lighter-range hydrocarbon (gasoline).

Economics

During the early 20th century, the crude-oil industry expanded. Since then, the various attempts to develop oil shale deposits have succeeded only when the cost of shale-oil production in a given region comes in below the price of crude oil or its other substitutes. According to a survey conducted by the RAND Corporation, the cost of producing a barrel of oil at a surface retorting complex in the United States (comprising a mine, retorting plant, upgrading plant, supporting utilities, and spent shale reclamation), would range between US$70–95 ($440–600/m^3, adjusted to 2005 values). This estimate considers varying levels of kerogen quality and extraction efficiency. In order to run a profitable operation, the price of crude oil would need to remain above these levels. The analysis also discusses the expectation that processing costs would drop after the establishment of the complex. The hypothetical unit would see a cost reduction of 35–70% after producing its first 500 million barrels (79×10^6 m^3). Assuming an increase in output of 25 thousand barrels per day (4.0×10^3 m^3/d) during each year after the start of commercial production, RAND predicts the costs would decline to $35–48 per barrel ($220–300/m^3) within 12 years. After achieving the milestone of 1 billion barrels (160×10^6 m^3), its costs would decline further to $30–40 per barrel ($190–250/m^3). Some commentators compare the proposed American oil-shale industry to the Athabasca oil-sands industry (the latter enterprise generated over 1 million barrels (160,000 m^3) of oil per day in late 2007), stating that "the first-generation facility is the hardest, both technically and economically".

Royal Dutch Shell has announced that its *in situ* extraction technology in Colorado could become competitive at prices over $30 per barrel ($190/m^3), while other technologies at full-scale production assert profitability at oil prices even lower than $20 per barrel ($130/m^3). To increase efficiency when retorting oil shale, researchers have proposed and tested several co-pyrolysis processes.

A 1972 publication in the journal *Pétrole Informations* compared shale-based oil production unfavorably with the coal liquefaction. The article portrayed coal liquefaction as less expensive, generating more oil, and creating fewer environmental impacts than extraction from oil shale. It cited a conversion ration of 650 litres (170 U.S. gal; 140 imp gal) of oil per one ton of coal, as against 150 litres (40 U.S. gal; 33 imp gal) of shale oil per one ton of oil shale.

A critical measure of the viability of oil shale as an energy source lies in the ratio of the energy produced by the shale to the energy used in its mining and processing, a ratio known as "Energy Returned on Energy Invested" (EROEI). A 1984 study estimated the EROEI of the various known oil-shale deposits as varying between 0.7–13.3 although known oil-shale extraction development projects assert an EROEI between 3 to 10. Royal Dutch Shell has reported an EROEI of three to four on its *in situ* development, Mahogany Research Project. The water needed in the oil shale retorting process offers an additional economic consideration: this may pose a problem in areas with water scarcity.

Environmental Considerations

Mining oil shale involves a number of environmental impacts, more pronounced in surface mining than in underground mining. They include acid drainage induced by the sudden rapid exposure and subsequent oxidation of formerly buried materials, the introduction of metals including mercury into surface-water and groundwater, increased erosion, sulfur-gas emissions, and air pollution caused by the production of particulates during processing, transport, and support activities. In 2002, about 97% of air pollution, 86% of total waste and 23% of water pollution in Estonia came from the power industry, which uses oil shale as the main resource for its power production.

Oil-shale extraction can damage the biological and recreational value of land and the ecosystem in the mining area. Combustion and thermal processing generate waste material. In addition, the atmospheric emissions from oil shale processing and combustion include carbon dioxide, a greenhouse gas. Environmentalists oppose production and usage of oil shale, as it creates even more greenhouse gases than conventional fossil fuels. Section 526 of the *Energy Independence And Security Act* prohibits United States government agencies from buying oil produced by processes that produce more greenhouse gas emissions than would traditional petroleum. Experimental *in situ* conversion processes and carbon capture and storage technologies may reduce some of these concerns in the future, but at the same time they may cause other problems, including groundwater pollution. Among the water contaminants commonly associated with oil shale processing are oxygen and nitrogen heterocyclic hydrocarbons. Commonly detected examples include quinoline derivatives, pyridine, and various alkyl homologues of pyridine (picoline, lutidine).

Some commentators have expressed concerns over the oil shale industry's use of water. In 2002, the oil shale-fired power industry used 91% of the water consumed in Estonia. Depending on technology, above-ground retorting uses between one and five barrels of water per barrel of produced shale-oil. A 2008 programmatic environmental impact statement issued by the US Bureau of Land Management stated that surface mining and retort operations produce 2 to 10 U.S. gallons (7.6 to 38 l; 1.7 to 8.3 imp gal) of waste water per 1 short ton (0.91 t) of processed oil shale. *In situ* processing, according to one estimate, uses about one-tenth as much water.

Water concerns become particularly sensitive issues in arid regions, such as the western US and Israel's Negev Desert, where plans exist to expand oil-shale extraction despite a water shortage.

Environmental activists, including members of Greenpeace, have organized strong protests against the oil shale industry. In one result, Queensland Energy Resources put the proposed Stuart Oil Shale Project in Australia on hold in 2004.

Enhanced Oil Recovery

Enhanced Oil Recovery (abbreviated EOR) is a generic term for techniques for increasing the amount of crude oil that can be extracted from an oil field. Using EOR, 30-60 %, or more, of the reservoir's original oil can be extracted compared with 20-40% using primary and secondary recovery.

Enhanced oil recovery is also called improved oil recovery or tertiary recovery (as opposed to primary and secondary recovery). Sometimes the term quaternary recovery is used to refer to more advanced, speculative, EOR techniques.

How does it Work?

Enhanced oil recovery is achieved by gas injection, chemical injection, microbial injection, or thermal recovery (which includes cyclic or continous steam, steam flooding, and fire flooding).

Gas Injection

Gas reinjection is presently the most-commonly used approach to enhanced recovery. In addition to the beneficial effect of the pressure, this method sometimes aids recovery by reducing the viscosity of the crude oil as the gas mixes with it.

Gases used include CO_2, natural gas or nitrogen. Air cannot be used to repressurize the reservoir because the oil will quickly catch on fire. Oil displacement by carbon dioxide injection relies on the phase behaviour of the mixtures of that gas and the crude, which are strongly dependent on reservoir temperature, pressure and crude oil composition. These mechanisms range from oil swelling and viscosity reduction for injection of immiscible fluids (at low pressures) to completely miscible displacement in high-pressure applications. In these applications, more than half and up to two-thirds of the injected CO_2 returns with the produced oil and is usually re-injected into the reservoir to minimize operating costs. The remainder is trapped in the oil reservoir by various means.

Chemical Injection

The injection of various chemicals, usually as dilute solutions, have been used to improve oil recovery. Injection of alkaline or caustic solutions into reservoirs with oil that has organic acids naturally occurring in the oil will result in the production of soap that may lower the interfacial tension enough to increase production. Injection of a dilute solution of a water soluble polymer to increase the viscosity of the injected water can increase the amount of oil recovered in some formations. Dilute solutions of surfactants such as petroleum sulfonates or biosurfactants such as rhamnolipids may be injected to lower the interfacial tension or capillary pressure that impedes oil droplets from moving through a reservoir. Special formulations of oil water and surfactant, microemulsions, can be particularly effective in this. Application of these methods is usually limited by the cost of the chemicals and their adsorption and loss onto the rock of the oil containing formation. In all of these methods the chemicals are injected into several wells and the production occurs in other nearby wells.

Microbial Injection

Microbial injection is part of microbial enhanced oil recovery and is presently rarely used, both because of its higher cost and because the developments in this field are more recent than other techniques. Strains of microbes have been both discovered and developed (using gene mutation) which function either by partially digesting long hydrocarbon molecules, by generating biosurfactants, or by emitting carbon dioxide (which then functions as described in Gas injection above).

Three approaches have been used to achieve microbial injection. In the first approach, bacterial cultures mixed with a food source (a carbohydrate such as molasses is commonly used) are injected into the oil field. In the second approach, used since 1985, nutrients are injected into the ground to nurture existing microbial bodies; these nutrients cause the bacteria to increase production of the natural surfactants they normally use to metabolize crude oil underground. After the injected nutrients are consumed, the microbes go into near-shutdown mode, their exteriors become hydrophilic, and they migrate to the oil-water interface area, where they cause oil droplets to form from the larger oil mass, making the droplets more likely to migrate to the wellhead. This approach has been used in oilfields near the Four Corners and in the Beverly Hills Oil Field in Beverly Hills, California.

The third approach is used to address the problem of paraffin components of the crude oil, which tend to separate from the crude as it flows to the surface. Since the Earth's surface is considerably cooler than the petroleum deposits (a temperature drop of 13-14 degree F per thousand feet of depth is usual), the paraffin's higher melting point causes it to solidify as it is cooled during the upward flow. Bacteria capable of breaking these paraffin chains into smaller chains (which would then flow more easily) are injected into the wellhead, either near the point of first congealment or in the rock stratum itself.

Thermal Methods

In this approach, various methods are used to heat the crude oil in the formation to reduce its viscosity and/or vaporize part of the oil. Methods include cyclic steam injection, steam drive and in situ combustion. These methods improve the sweep efficiency and the displacement efficiency. Steam injection has been used commercially since the 1960s in California fields. In 2011 Solar Thermal Enhanced Oil Recovery projects were started in California and Oman, this method is similar to Thermal EOR but uses a solar array to produce the steam.

Economic Costs and Benefits

Adding oil recovery methods adds to the cost of oil — in the case of CO_2 typically between 0.5-8.0 US$ per tonne of CO_2. The increased extraction of oil on the other hand, is an economic benefit with the revenue depending on prevailing oil prices. Onshore EOR has paid in the range of a net 10-16 US$ per tonne of CO_2 injected for oil prices of 15-20 US$/barrel. Prevailing prices depend on many factors but can determine the economic suitability of any procedure, with more procedures and more expensive procedures being economically viable at higher prices. Example: With oil prices at around 90 US$/barrel, the economic benefit is about 70 US$ per tonne CO_2.

Examples of Current EOR Projects

In Canada, a CO_2-EOR project has been established by Cenovus Energy at the Weyburn Oil Field in southern Saskatchewan. The project is expected to inject a net 18 million ton CO_2 and recover an additional 130 million barrels (21,000,000 m^3) of oil, extending the life of the oil field by 25 years. There is a projected 26+ million tonnes (net of production) of CO2 to be stored in Weyburn, plus another 8.5 million tonnes (net of production) stored at the Weyburn-Midale Carbon Dioxide Project, resulting in a net reduction in atmospheric CO_2). That's the equivalent of taking nearly 7 million cars off the road for a year. Since CO_2 injection began in late 2000, the EOR project has performed largely as predicted. Currently, some 1600 m^3 (10,063 barrels) per day of incremental oil is being produced from the field.

Potential for EOR in United States

In United States, the Department of Energy (DOE) has estimated that full use of 'next generation' CO_2-EOR in United States could generate an additional 240 billion barrels (38 km^3) of recoverable oil resources. Developing this potential would depend on the availability of commercial CO_2 in large volumes, which could be made possible by widespread use of carbon capture and storage. For comparison, the total undeveloped US domestic oil resources still in the ground total more than 1 trillion barrels (160 km^3), most of it remaining unrecoverable. The DOE estimates that if the EOR potential were to be fully realised, State and local treasuries would gain $280 billion in revenues from future royalties, severance taxes, and state income taxes on oil production, aside from other economic benefits.

Environmental Impacts

Enhanced oil recovery wells typically produce large quantities of brine at the surface. The brine may contain toxic metals and radioactive substances, as well as being very salty. This can be very damaging to drinking water sources and the environment generally if not properly controlled. In the United States, injection well activity is regulated by the United States Environmental Protection Agency (EPA) and state governments under the Safe Drinking Water Act. EPA has issued Underground Injection Control (UIC) regulations in order to protect drinking water sources. The regulations require well operators to reinject the brine deep underground.

Petroleum Engineering

Petroleum engineering is an engineering discipline concerned with the activities related to the production of hydrocarbons, which can be either crude oil or natural gas. Subsurface activities are deemed to fall within the *upstream* sector of the oil and gas industry, which are the activities of finding and producing hydrocarbons.

Refining and distribution to a market are referred to as the *downstream* sector. Exploration, by earth scientists, and petroleum engineering are the oil and gas industry's two main subsurface disciplines, which focus on maximizing economic recovery of hydrocarbons from subsurface reservoirs. Petroleum geology and geophysics focus on provision of a static description of the hydrocarbon reservoir rock, while petroleum engineering focuses on estimation of the recoverable volume of this resource using a detailed understanding of the physical behavior of oil, water and gas within porous rock at very high pressure.

The combined efforts of geologists and petroleum engineers throughout the life of a hydrocarbon accumulation determine the way in which a reservoir is developed and depleted, and usually they have the highest impact on field economics. Petroleum engineering requires a good knowledge of many other related disciplines, such as geophysics, petroleum geology, formation evaluation (well logging), drilling, economics, reservoir simulation, well engineering, artificial lift systems, and oil and gas facilities engineering.

Petroleum engineering has become a technical profession that involves extracting oil in increasingly difficult situations as much of the "low hanging fruit" of the world's oil fields has been found and depleted. Improvements in computer modelling, materials and the application of statistics, probability analysis, and new technologies like horizontal drilling and enhanced oil recovery, have drastically improved the toolbox of the petroleum engineer in recent decades.

Deep-water, arctic and desert conditions are commonly contended with. High Temperature and High Pressure (HTHP) environments have become increasingly commonplace in operations and require the petroleum engineer to be savvy in topics as wide ranging as thermo-hydraulics, geomechanics, and intelligent systems.

The Society of Petroleum Engineers (SPE) is the largest professional society for petroleum engineers and publishes much information concerning the industry. Petroleum engineering education is available at 17 universities in the United States and many more throughout the world - primarily in oil producing regions - and some oil companies have considerable in-house petroleum engineering training classes.

Petroleum engineering has historically been one of the highest paid engineering disciplines, although there is a tendency for mass layoffs when oil prices decline. In a June 4th, 2007 article, Forbes.com reported that petroleum engineering was the 24th best paying job in the United States. The 2010 National Association of Colleges and Employers survey showed petroleum engineers as the highest paid 2010 graduates at an average $125,220 annual salary. For individuals with experience, salaries can go from $170,000 to $260,000 annually.

Types

Petroleum engineers divide themselves into several types:

- Reservoir engineers work to optimize production of oil and gas via proper well placement, production rates, and enhanced oil recovery techniques.
- Drilling engineers manage the technical aspects of drilling exploratory, production and injection wells.

Society of Petroleum Engineers

The Society of Petroleum Engineers (SPE) is a not-for-profit professional organization whose mission is to collect, disseminate, and exchange technical knowledge concerning the exploration, development and production of oil and gas resources and related technologies for the public benefit and to provide opportunities for professionals to enhance their technical and professional competence.

The SPE provides a worldwide forum of oil and natural gas exploration and production (E&P) professionals for the exchange of technical knowledge and a professional home for more than 88,000 engineers, scientists, managers, and educators. SPE's technical library contains more than 42,000 technical papers — products of SPE conferences and periodicals, made available to the entire industry. SPE has offices in Dallas, Houston, Calgary, London, Dubai, Moscow and Kuala Lumpur.

History

The history of the SPE began well before its actual establishment. During the decade after the 1901 discovery of the Spindletop field, the American Institute of Mining Engineers (AIME) saw a growing need for a forum in the booming new field of petroleum engineering. As a result, AIME formed a standing committee on oil and gas in 1913.

In 1922, the committee was expanded to become one of AIME's 10 professional divisions. The Petroleum Division of AIME continued to grow throughout the next three decades. By 1950, the Petroleum Division had become one of three separate branches of AIME, and in 1957 the Petroleum Branch of AIME was expanded once again to form a professional society.

The first SPE Board of Directors meeting was held 6 October 1957, making 2007 the 50th anniversary year for SPE as a professional society.

Chronology

- 1950s: During the 1950s, the petroleum membership of AIME grew significantly, leading to restructuring decisions that would shape the future Society of Petroleum Engineers.
- 1957: The Petroleum Branch of AIME becomes a full-fledged professional society - the Society of Petroleum Engineers of AIME. On October 6, 1957, the first Board of Directors meeting was held in Dallas, Texas, with President John H. Hammond presiding.
- 1958: The SPE Reprint Series begins with the publication of Well Logging.
- 1961: The first issue of the Society of Petroleum Engineers Journal is published.
- 1979: The first Middle East Oil and Gas Show and Conference is held.
- 1985: SPE is incorporated separately from AIME.

OnePetro

Launched in March 2007, OnePetro.org is a multi-society library that allows users to search for and access a broad range of technical literature related to the oil and gas exploration and production industry. OnePetro is a multi-association effort that reflects participation of many organizations. The Society of Petroleum Engineers (SPE) operates OnePetro on behalf of the participating organizations. SPE provides the computers and technology on which OnePetro operates and provides customer service support.

OnePetro currently contains more than 105,000 documents, with more being added frequently. The number of documents is expected to grow as additional organizations choose to make their materials available through OnePetro. OnePetro is the first online offering of documents from some organizations, making these materials widely available for the first time.

The following organizations currently have their technical documents available through OnePetro:

- American Petroleum Institute (API)
- American Rock Mechanics Association (ARMA)
- American Society of Safety Engineers (ASSE)
- Offshore Technology Conference (OTC)
- NACE International (corrosion engineers)
- Petroleum Society of Canada (PETSOC)
- Society of Petroleum Engineers (SPE)
- Society of Petrophysicists and Well Log Analysts (SPWLA)
- Society for Underwater Technology (SUT)
- World Petroleum Council (WPC).

SPE Petroleum Engineering Certification

The SPE Petroleum Engineering Certification program was instituted as a way to certify petroleum engineers by examination and experience. This certification is similar to the Registration of Petroleum Engineers by the States in the United States.

Certified professionals use SPEC after their name

Petroleum Reserves and Resources Definitions

The Society of Petroleum Engineers has developed a system for evaluating oil and gas reserves and resources. The Petroleum Resources

Management System (PRMS) is used by oil and gas companies in determining their reserves and serves as the primary basis for reporting rules established by the United States Securities and Exchange Commission. Petroleum Resources Management System

- equipment that separates the produced fluids (oil, natural gas, and water).

Seismic to Simulation

Seismic to Simulation is the process and associated techniques used to develop highly accurate static and dynamic 3D models of hydrocarbon reservoirs for use in predicting future production, placing additional wells, and evaluating alternative reservoir management scenarios. The process is successful if the model accurately reflects the original well logs, seismic data and production history.

Introduction

Reservoir models are constructed to gain a better understanding of the subsurface that leads to informed well placement, reserves estimation and production planning. Models are based on measurements taken in the field, including well logs, seismic surveys, and production history.

Seismic to simulation enables the quantitative integration of all field data into an updateable reservoir model built by a team of geologists, geophysicists, and engineers. Key techniques used in the process include integrated petrophysics and rock physics to determine the range of lithotypes and rock properties, geostatistical inversion to determine a set of plausible seismic-derived rock property models at sufficient vertical resolution and heterogeneity for flow simulation, stratigraphic grid transfer to accurately move seismic-derived data to the geologic model, and flow simulation for model validation and ranking to determine the model that best fits all the data.

Rock Physics and Petrophysics

The first step in seismic to simulation is establishing a relationship between petrophysical key rock properties and elastic properties of the rock. This is required in order to find common ground between the well logs and seismic data.

Well logs are measured in depth and provide high resolution vertical data, but no insight into the inter-well space. Seismic are measured in time and provide great lateral detail but is quite limited in its vertical resolution. When correlated, well logs and seismic can be used to create a fine-scale 3D model of the subsurface.

Insight into the rock properties comes from a combination of basic geologic understanding and well bore measurements. Based on an understanding of how the area was formed over time, geologists can predict the types of rock likely to be present and how rapidly they vary spatially. Well log and core measurements provide samples to verify and fine-tune that understanding.

Seismic data is used by petrophysicists to identify the tops of various lithotypes and the distribution of rock properties in the inter-well space using seismic inversion attributes such as impedance. Seismic surveys measure acoustic impedance contrasts between rock layers. As different geologic structures are encountered, the sound wave reflects and refracts as a function of the impedance contrast between the layers. Acoustic impedance varies by rock type and can therefore be correlated to rock properties using rock physics relationships between the inversion attributes and petrophysical properties such as porosity, lithology, water saturation, and permeability.

Once well logs are properly conditioned and edited, a petrophysical rock model is generated that can be used to derive the effective elastic rock properties from fluid and mineral parameters as well as rock structure information. The model parameters are calibrated by comparison of the synthetic to the available elastic sonic logs. Calculations are performed following a number of rock physics algorithms including: Xu & White, Greenberg & Castagna, Gassmann, Gardner, modified upper and lower Hashin-Shtrikman, and Batzle & Wang.

When the petrophysical rock model is complete, a statistical database is created to describe the rock types and their known properties such as porosity and permeability. Lithotypes are described, along with their distinct elastic properties.

MCMC Geostatistical Inversion

In the next step of seismic to simulation, seismic inversion techniques combine well and seismic data to produce multiple equally plausible 3D models of the elastic properties of the reservoir. Seismic data is transformed to elastic property log(s) at every trace. Deterministic inversion techniques are used to provide a good overall view of the porosity over the field, and serve as a quality control check. To obtain greater detail needed for complex geology, additional stochastic inversion is then employed.

Geostatistical inversion procedures detect and delineate thin reservoirs otherwise poorly defined. Markov chain Monte Carlo (MCMC) based geostatistical inversion addresses the vertical scaling problem by creating seismic derived rock properties with vertical sampling compatible to geologic models. All field data is incorporated into the geostatistical inversion process through the use of probability distribution functions (PDFs). Each PDF describes a particular input data in geostatistical terms using histograms and variograms, which identify the odds of a given value at a specific place and the overall expected scale and texture based on geologic insight.

Once constructed, the PDFs are combined using Bayesian inference, resulting in a posterior PDF that conforms to everything that is known about the field. A weighting system is used within the algorithm, making the process more objective.

From the posterior PDF, realizations are generated using a Markov chain Monte Carlo algorithm. These realizations are statistically fair and produce models of high detail, accuracy and realism. Rock properties like porosity can be cosimulated from the elastic properties determined by the geostatistical inversion. This process is iterated until a best fit model is identified.

Inversion parameters are tuned by running the inversion many times with and without well data. Without the well data, the inversions are running in blind-well mode. These blind-well mode inversions test the reliability of the constrained inversion and remove potential biased.

This statistical approach creates multiple, equi-probable models consistent with the seismic, wells, and geology. Geostatistical inversion simultaneously inverts for impedance and discrete properties types, and other petrophysical properties such as porosity can then be jointly cosimulated.

The output volumes are at a sample rate consistent with the reservoir model because making synthetics of finely sampled models is the same as from well logs. Inversion properties are consistent with well log properties because the histograms used to generate the output rock properties from the inversion are based on well log values for those rock properties.

Uncertainty is quantified by using random seeds to generate slightly differing realizations, particularly for areas of interest. This process improves the understanding of uncertainty and risk within the model.

Stratigraphic Grid Transfer

Following geostatistical inversion and in preparation for history matching and flow simulation, the static model is re-gridded and up-scaled. The transfer simultaneously converts time to depth for the various properties and transfers them in 3D from the seismic grid to a corner-point grid. The relative locations of properties are preserved, ensuring data points in the seismic grid arrive in the correct stratigraphic layer in the corner point grid.

The static model built from seismic is typically orthogonal but flow simulators expect corner point grids. The corner point grid consists of cubes that are usually much coarser in the horizontal direction and each corner of the cube is arbitrarily defined to follow the major features in the grid. Converting directly from orthogonal to corner point can cause problems such as creating discontinuity in fluid flow.

An intermediate stratigraphic grid ensures that important structures are not misrepresented in the transfer. The stratigraphic grid has the same number of cells as the orthogonal seismic grid, but the boundaries are defined by stratigraphic surfaces and the cells follow the stratigraphic organization. This is a stratigraphic representation of the seismic data using the seismic interpretation to define the layers. The stratigraphic grid model is then mapped to the corner point grid by adjusting the zones.

Using the porosity and permeability models and a saturation height function, initial saturation models are built. If volumetric calculations identify problems in the model, changes are made in the petrophysical model without causing the model to stray from the original input data. For example, sealing faults are added for greater compartmentalization.

Model Validation and Ranking

In the last step of seismic to simulation, flow simulation continues the integration process by bringing in the production history. This provides a further validation of the static model against history. A representative set of the model realizations from the geostatistical inversion are history matched against production data. If the properties in the model are realistic, simulated well bottom hole pressure behavior should match historical (measured) well bottom hole pressure. Production flow rates and other engineering data should also match.

Based on the quality of the match, some models are eliminated. After the initial history match process, dynamic well parameters are adjusted as needed for each of the remaining models to improve the match. The final model represents the best match to original field measurements and production data and is then used in drilling decisions and production planning.

Microbial Enhanced Oil Recovery

Microbial Enhanced Oil Recovery (MEOR) is a biological based technology consisting in manipulating function or structure, or both, of microbial environments existing in oil reservoirs. The ultimate aim of MEOR is to improve the recovery of oil entrapped in porous media while increasing economic profits. MEOR is a tertiary oil extraction technology allowing the partial recovery of the commonly residual two-thirds of oil, thus increasing the life of mature oil reservoirs.

MEOR is a multidisciplinary field incorporating, among others: geology, chemistry, microbiology, fluids mechanics, petroleum engineering, environmental engineering and chemical engineering. The microbial processes proceeding in MEOR can be classified according to the oil production problem in the field:

- *well bore clean up* removes mud and other debris blocking the channels where oil flows through;
- *well stimulation* improves the flow of oil from the drainage area into the well bore; and
- *enhanced water floods* increase microbial activity by injecting selected microbes and sometimes nutrients. From the engineering point of view, MEOR is a system integrated by the reservoir, microbes, nutrients and protocol of well injection.

MEOR Outcomes

So far, the outcomes of MEOR are explained based on two predominant rationales:

Increment in oil production. This is done by modifying the interfacial properties of the system oil-water-minerals, with the aim of facilitating oil movement through porous media. In such a system, microbial activity affects fluidity (viscosity reduction, miscible flooding);

displacement efficiency (decrease of interfacial tension, increase of permeability); sweep efficiency (mobility control, selective plugging) and driving force (reservoir pressure).

Upgrading. In this case, microbial activity acts may promote the degradation of heavy oils into lighter ones. Alternatively, it can promote desulphurization due to denitrification as well as the removal of heavy metals.

Relevance

Several decades of research and successful applications support the claims of MEOR as a mature technology. Despite those facts, disagreement still exists. Successful stories are specific for each MEOR field application, and published information regarding supportive economical advantages is however inexistent. Despite this, there is consensus considering MEOR one of the cheapest existing EOR methods. However, obscurity exists on predicting whether or not the deployment of MEOR will be successful. MEOR is, therefore, one of the future research areas with great priority as identified by the "Oil and Gas in the 21st Century Task Force". This is probably because MEOR is a complementary technology that may help recover the 377 billion barrels of oil that are unrecoverable by conventional technologies.

Bias

Before the advent of environmental molecular microbiology, the word "bacteria" was utilised indistinctively in many fields to refer to uncharacterized microbes, and such systematic error affected several disciplines. Therefore, the word "microbe" or "microorganism" will therefore be preferred hereafter in the text.

History

It was in 1926 when Beckam proposed the utilisation of microorganisms as agents for recovering the remnant oil entrapped in porous media. Since that time numerous investigations have been developed, and are extensively reviewed. In 1947, ZoBell and colleagues set the basis of petroleum microbiology applied to oil recovery, whose contribution would be useful for the first MEOR patent granted to Updegraff and colleagues in 1957 concerning the in situ production of oil recovery agents such as gases, acids, solvents and biosurfactants from microbial degradation of molasses. In 1954, the first field test was carried out in the Lisbon field in Arkansas, USA. During that time, Kuznetsov discovered the microbial gas production from oil. From this year and until the 1970s there was intensive research in USA, USSR, Czechoslovakia, Hungary and Poland. The main type of field experiments developed in those countries consisted in injecting exogenous microbes.

In 1958, selective plugging with microbial produced biomass was proposed by Heinningen and colleagues. The oil crisis of 1970 triggered a great interest in active MEOR research in more than 15 countries. From 1970 to 2000, basic MEOR research focused on microbial ecology and characterization of oil reservoirs. In 1983, Ivanov and colleagues developed the strata microbial activation technology. By 1990, MEOR achieved an interdisciplinary technology status. In 1995, a survey of MEOR projects (322) in the USA showed that 81% of the projects successfully increased oil production, and there was not a single case of reduced oil production. Today, MEOR is gaining attention owing to the high prices of oil and the imminent ending of this resource. As a result, several countries are willing to use MEOR in one third of their oil recovery programs by 2010.

MEOR Advantages

There is a plethora of reviewed claims regarding the advantages of MEOR. However, they should be cautiously regarded due to the lack of published supportive evidence. In addition, assessments of both full live cycle analysis and environmental impact are also unknown.

Advantages can be summarised as follows:

- Injected microbes and nutrients are cheap; easy to handle in the field and independent of oil prices.
- Economically attractive for mature oil fields before abandonment.
- Increases oil production.
- Existing facilities require slight modifications.
- Easy application.
- Less expensive set up.
- Low energy input requirement for microbes to produce MEOR agents.
- More efficient than other EOR methods when applied to carbonate oil reservoirs.

- Microbial activity increases with microbial growth. This is opposite to the case of other EOR additives in time and distance.
- Cellular products are biodegradable and therefore can be considered environmentally friendly.

MEOR Disadvantages

- The oxygen deployed in aerobic MEOR can act as corrosive agent on non-resistant topside equipment and down-hole piping
- Anaerobic MEOR requires large amounts of sugar limiting its applicability in offshore platforms due to logistical problems
- Exogenous microbes require facilities for their cultivation.
- Indigenous microbes need a standardized framework for evaluating microbial activity, e.g. specialized coring and sampling techniques.
- Microbial growth is favoured when: layer permeability is greater than 50 md; reservoir temperature is inferior to 80^{0}C, salinity is below 150 g/L and reservoir depth is less than 2400m.

The Environment of an Oil Reservoir

Oil reservoirs are complex environments containing living (microorganisms) and non living factors (minerals) which interact with each other in a complicated dynamic network of nutrients and energy fluxes. Since the reservoir is heterogeneous, so do the variety of ecosystems containing diverse microbial communities, which in turn are able to affect reservoir behaviour and oil mobilization. Microbes are living machines whose metabolites, excretion products and new cells may interact with each other or with the environment, positively or negatively, depending on the global desirable purpose, e.g. the enhancement of oil recovery. All these entities, i.e. enzymes, extracellular polymeric substances (EPS) and the cells themselves, may participate as catalyst or reactants. Such complexity is increased by the interplay with the environment, the later playing a crucial role by affecting cellular function, i.e. genetic expression and protein production. Despite this fundamental knowledge on cell physiology, a solid understanding on function and structure of microbial communities in oil reservoirs, i.e. ecophysiology, remains inexistent.

Environmental Constraints

Several factors concomitantly affect microbial growth and activity. In oil reservoirs, such environmental constraints permit the establishment of criteria to assess and compare the suitability of various microorganisms. Those constraints may not be as harsh as other environments on Earth. For example with, connate brines the salinity is higher than that of sea water but lower than that of salt lakes. In addition, pressures up to 20 MPa and temperatures up to 80 °C, in oil reservoirs, are within the limits for the survival of other microorganisms. Some environmental constraints creating selective pressures on cellular systems that may also affect microbial communities in oil reservoirs are:

Temperature

Enzymes are biological catalysts whose function is affected by a variety of factors including temperature, which, at different ranges, may improve or hamper enzymatic mediated reactions. This will have an effect over the optimal cellular growth or metabolism. Such dependency permits classification of microbes according to the range of temperatures at which they grow. For instance: psychrophiles (<25 °C), mesophiles (25-45 °C), thermophiles (45-60 °C) and hyperthermophiles (60-121 °C). Although such cells optimally grow in these temperature ranges there may not be a direct relationship with the production of specific metabolites.

Direct Effects

The effects of pressure on microbial growth under deep ocean conditions were investigated by ZoBell and Johson in 1949. They called those microbes whose growth was enhanced by increasing pressure, barophilic. Other classifications of microorganisms are based on whether microbial growth is inhibited at standard conditions (piezophiles) or above 40 MPa (piezotolerants). From a molecular point of view, the review of Daniel shows that at high pressures the DNA double helix becomes denser, and therefore both gene expression and protein synthesis are affected.

Indirect Effect

Increasing pressure increases gas solubility, and this may affect the redox potential of gases participating as electron acceptors and donors, such as hydrogen or CO_2.

Pore Size/Geometry

One study has concluded that substantial bacterial activity is achieved when there are interconnections of pores having at least 0.2μ diameter. It is expected that pore size and geometry may affect chemotaxis. However, this has not been proven at oil reservoir conditions.

pH

The acidity of alkalinity has an impact over several aspects in living and non living systems. For instance:

Surface Charge

Changes in cellular surface and membrane thickness may be promoted by pH due to its ionization power of cellular membrane embedded proteins. The modified ionic regions may interact with mineral particles and affect the motion of cells through the porous media.

Enzymatic Activity

Embedded cell proteins play a fundamental roll in the transport of chemicals across the cellular membrane. Their function is strongly dependent on their state of ionisation, which is in turn strongly affected by pH.

In both cases, this may happen in isolated or complex environmental microbial communities. So far the understanding on the interaction between pH and environmental microbial communities remains unknown, despite the efforts of the last decade. Little is know on the ecophysiology of complex microbial communities and research is still in developmental stage..

Oxidation Potential

The oxidation potential (Eh, measured in volts) is, as in any reaction system, the thermodynamic driving force of anaerobic respiration, which takes place in oxygen depleted environments. Prokaryotes are among the cells that have anaerobic respiration as metabolic strategy for survival. The electron transport takes place along and across the cellular membrane (prokaryotes lack of mitochondria). Electrons are transferred from an electron donor (molecule to be oxidised anaerobically) to an electron acceptor (NO_3, SO_4, MnO_4, etc.). The net Eh between a given electron donor and acceptor; hydrogen ions and other species in place will determine which reaction will first take place. For instance, nitrification is hierarchically more favoured than sulphate reduction. This allows for enhanced oil recovery by disfavouring biologically produced H_2S, which derives from reduced SO_4. In this process, the effects of nitrate reduction on wettability, interfacial tension, viscosity, permeability, biomass and biopolymer production remain unknown.

Electrolyte Composition

Electrolytes concentration and other dissolved species may affect cellular physiology. Dissolving electrolytes reduces thermodynamic activity (aw), vapour pressure and autoprotolysis of water. Besides, electrolytes promote an ionic strength gradient across cellular membrane and therefore provides a powerful driving force allowing the diffusion of water into or out to cells. In natural environments, most bacteria are incapable of living at aw below 0.95. However, some microbes from hypersaline environment such as Pseudomonas species and Halococcus thrive at lower a_w, and are therefore interesting for MEOR research.

Non-specific Effects

They may occur on pH and Eh. For example, increasing ionic strength increases solubility of nonelectrolytes ('salting out') as in the case of dissolution of carbon dioxide, a pH controller of a variety of natural waters.

Biological Factors

Although it is widely accepted that predation, parasitism, syntrophism and other relationships also occur in the microbial world, little is known in this relationships on MEOR and they have been disregarded in MEOR experiments. In other cases, some microorganisms can thrive in nutrient deficient environments (oligotrophy) such as deep granitic and basaltic aquifers. Other microbes, living in sediments, may utilise available organic compounds (heterotrophy). Organic matter and metabolic products between geological formations can diffuse and support microbial growth in distant environments.

MEOR Mechanism

Understanding MEOR mechanism is still far from being clear. Although a variety of explanations has been given in isolated experiments, it is unclear if they were carried out trying to mimic oil reservoirs conditions.

The mechanism can be explained from the client-operator viewpoint which considers a series of concomitant positive or negative effects that will result in a global benefit:

- *Beneficial effects.* Biodegradation of big molecules reduces viscosity; production of surfactants reduces interfacial tension; production of gas provides additional pressure driving force; microbial metabolites or the microbes themselves may reduce permeability by activation of secondary flow paths.
- *Detrimental effects.* Biologically produced hydrogen sulphide, i.e. souring, causes corrosion of piping and machinery; consumption of hydrocarbons by bacteria reduces the production of desired chemicals.
- *Beneficial or Detrimental.* Permeability reduction can be beneficial in some cases but detrimental in others. Negatively, microbial metabolites or the microbes themselves may reduce permeability by activation of secondary flow paths by depositing: biomass (biological clogging), minerals (chemical clogging) or other suspended particles (physical clogging). Positively, attachment of bacteria and development of slime, i.e. extracellular polymeric substances (EPS), favour the plugging of highly permeable zones (thieves zones) leading to increased sweep efficiency.

MEOR Strategies

Changing oil reservoir ecophysiology to favour MEOR can be achieved by complementing different strategies. In situ microbial stimulation can be chemically promoted by injecting electron acceptors such as nitrate; easy fermentable molasses, vitamins or surfactants. Alternatively, MEOR is promoted by injecting exogenous microbes, which may be adapted to oil reservoir conditions and be capable of producing desired MEOR agents.

This knowledge has been obtained from experiments with pure cultures and some times with complex microbial communities but the experimental conditions are far from mimicking those ones prevailing in oil reservoirs. It is unknown if metabolic products is cell growth dependent, and claims in this respect should be taken cautiously, since the production of a metabolite is not always dependent of cellular growth.

Biomass and Biopolymers

In selective plugging, conditioned cells and extracellular polymeric substances plug high permeability zones, resulting in a change of direction of the water flood to oil-rich channels, consequently increasing the sweep efficiency of oil recovery with water flooding. Biopolymer production and the resulting biofilm formation (less 27% cells, 73-98% EPS and void space) are affected by water chemistry, pH, surface charge, microbial physiology, nutrients and fluid flow.

Biosurfactants

Microbial produced surfactants, i.e. biosurfactants reduce the interfacial tension between water and oil, and therefore a lower hydrostatic pressure is required to move the liquid entrapped in the pores to overcome the capillary effect.

Secondly, biosurfactants contribute to the formation of micelles providing a physical mechanism to mobilise oil in a moving aqueous phase. Hydrophobic and hydrophilic compounds are in play and have attracted attention in MEOR research, and the main structural types are lipopeptides and glycolipids, being the fatty acid molecule the hydrophobic part.

Gas and Solvents

In this old practice, the production of gas has a positive effect in oil recovery by increasing the differential pressure driving the oil movement. Anaerobically produced methane from oil degradation have a low effect on MEOR due to its high solubility at high pressures. Carbon dioxide is also a good MEOR agent. The miscible CO_2 is condensed into the liquid phase when light hydrocarbons are vaporised into the gas phase. Immiscible CO_2 helps to saturate oil, resulting in swelling and reduction of viscosity of the liquid phase and consequently improving mobilization by extra driving pressure. Concomitantly, other gases and solvents may dissolve carbonate rock, leading to an increase in rock permeability and porosity.

Field Studies

Worldwide MEOR field applications have been reviewed in detail. Although the exact number field trials is unknown, Lazar et al. suggested an order of hundreds. Successful MEOR field trials have been conducted in the U.S., Russia, China, Australia, Argentina, Bulgaria, former Czechoslovakia, former East Germany, Hungary, India, Malaysia, Peru, Poland, and Romania. Lazar et al. suggested China is leading in the area, and also found that the most successful study was carried out in Alton field, Australia (40% increase of oil production in 12 months).

The majority of the field trials were done in sandstone reservoirs and very few in fractured reservoirs and carbonates. The only known offshore field trials were in Norne (Norway) and Bokor (Malaysia).

As reviewed by Lazar et al., field application followed different approaches such as injection of exogenous microorganisms (microbial flooding); control of paraffin deposition; stimulation of indigenous microbes; injection of ex situ produced biopolymers; starved selected ultramicrobes (selected plugging); selected plugging by sand consolidation due to biomineralization and fracture clogging in carbonate formations; nutrient manipulation of indigenous reservoir microbes to produce ultramicrobes; and adapted mixed enrichment cultures.

Reported MEOR results from field trials vary widely. Rigorous controlled experiments are lacking and may not be possible due to the dynamic changes in the reservoir when oil is being recovered. Besides, the economical advantages of these field trials are unknown, and the answer to why the other trials were unsuccessful is unknown. General conclusions can not be drawn because the physical and mineralogical characteristics of the oil reservoirs reported were different. The extrapolation of such conclusions is therefore unviable.

Models

A plethora of attempts to model MEOR has been published. Until now, it is unclear if theoretical results reflect the scarce published data. Developing mathematical models for MEOR is very challenging since physical, chemical and biological factors need to be considered.

Published MEOR models are composed of transport properties, conservation laws, local equilibrium, breakdown of filtration theory and physical straining. Such models are so far simplistic and they were developed based on:

(A) Fundamental conservation laws, cellular growth, retention kinetics of biomass, and biomass in oil and aqueous phases. The main aim was to predict porosity retention as a function of distance and time.

(B) Filtration model to express bacterial transport as a function of pore size; and relate permeability with the rate of microbial penetration by applying Darcy's law.

Chemical kinetics is fundamental for coupling bioproduct formation to fluxes of aqueous species and suspended microbes. Fully numerical approaches have also been followed. For instance, coupled nonlinear parabolic differential equations: adding equation for the rate of diffusion of microbes and their capture by porous medium; differential balance equations for nutrient transport, including the effect of adsorption; and the assumption of bacterial growth kinetic based on Monod equation.

Monod equation is indiscriminately used in modelling software, and has a limited behaviour which is inconsistent with the law of mass action that form the basis of kinetic characterization of microbial growth. Application of law of mass action to microbial populations results in the linear logistic equation. And the application of the law of mass action to an enzyme-catalysed process results in the Michaelis-Menten equation, from which Monod is inspired. This makes things difficult for in situ biosurfactant production because controlled experimentation is required to determine specific growth rate and Michaelis-Menten parameters of rate-limiting enzyme reaction.

Modelling of bioclogging is complicated because the production of clogging metabolite is coupled nonlinearly to the growth of microbes and flux of nutrients transported in the fluid.

Published models disregard the ecophysiology of the entire microbial microcosms at oil reservoir conditions. Microorganisms are a kind of catalyst whose activity (physiology) depends on the mutual interplay with other microbes and the environment (ecology). In nature, living and non living elements interact with each other in a complicated network of nutrients and energy. Some microbes produce extracellular polymeric substances and therefore its behaviour in pours media needs to consider both occupation by the EPS and the microbes themselves. Knowledge is lacking in this respect and therefore the aim of maximizing yield and minimizing cost remains unachieved.

Realistic models for MEOR at the conditions of the oil reservoir are missing, and reported parallel-pore models had fundamental deficiencies that were overcome by models considering the clogging of pores by microbes or biofilms, but such models have also the deficiency of being two-dimensional. The utilisation of such models in three dimensional models has not been proven. It is uncertain if they can be incorporated to popular oilfield simulation software. Thus, a field strategy needs a simulator capable of predicting bacterial growth and transport through porous network and in situ production of MEOR agents.

Grounds of Failure

- Lack of holistic approach allowing for a critical evaluation of economics, applicability and performance of MEOR is missing.
- No published study includes reservoir characteristics; biochemical and physiological characteristics of microbiota; controlling mechanisms and process economics.
- The ecophysiology of microbial communities thriving in oil reservoirs is largely unexplored. Consequently, there is a poor critical evaluation of the physical and biochemical mechanisms controlling microbial response to the hydrocarbon substrates and their mobility.
- Absence of quantitative understanding of microbial activity and poor understanding of the synergistic interactions between living and none living elements. Experiments based on pure cultures or enrichments are questionable because microbial communities interact synergistically with minerals, extracellular polymeric substances and other physicochemical and biological factors in the environment.
- Lack of cooperation between microbiologists, reservoir engineers, geologists, economists and owner operators; incomplete pertinent reservoir data, in published sources: lithology, depth, net thickness, porosity, permeability, temperature, pressure, reserves, reservoir fluid properties (oil gravity, water salinity, oil viscosity, bubble point pressure, and oil-formation-volume factor), specific EOR data (number of production and injection wells, incremental recovery potential as mentioned by the operator, injection rate, calculated daily and total enhanced production), calculated incremental recovery potential over the reported time.
- Limited understanding of MEOR process economics and improper assessment of technical, logistical, cost, and oil recovery potential.
- Unknowns life cycle assessments. Unknown environmental impact
- Lack of demonstrable quantitative relationships between microbial performance, reservoir characteristics and operating conditions

- Inconsistency in situ performance; low ultimate oil recovery factor; uncertainty about meeting engineering design criteria by microbial process; and a general apprehension about process involving live bacteria.
- Lack of rigorous controlled experiments, which are far from mimicking oil reservoir conditions that may have an effect over gene expression and protein formation.
- Kinetic characterization of bacteria of interest is unknown. Monod equation has been broadly misused.
- Lack of structured mathematical models to better describe MEOR.
- Lack of understanding of microbial oil recovery mechanism and deficient mathematical models to predict microbial behaviour in different reservoirs.
- Surfactants: biodegradable, effectiveness affected by temperature, pH and salt concentration; adsorption on to rock surfaces.
- Unfeasible economic solutions such as the utilization of enzymes and cultured microorganism.
- Difficult isolation or engineering of good candidate strains able to survive the extreme environment of oil reservoirs (up to 85 °C, up to 17.23 MPa).

Trends

- Wellbore microbial plugging and consequent lost of injectivity (clogging).
- Dispersion of components necessary to the target.
- Control of indigenous microbial activity.
- Mitigation of unwanted secondary activity due to competitive redox processes such as sulphate reduction, i.e. control of souring.
- Microbial paraffin removal.
- Microbial skin damage removal.
- Water floods, where continuous water phase enables the introduction of MEOR.
- Single-well stimulation, here the low cost makes MEOR the best choice.

- Selective plugging strategies.
- MEOR with ultramicrobes.
- Genetically engineered MEOR microorganisms able to survive, grow and produce metabolites at the expense of cheap nutrients and substrates.
- Application of extremophiles: halophiles, barophiles, and thermophiles.
- Artificial neural network modelling for describing in situ MEOR processes.
- Competition of exogenous microbes with indigenous micro flora, no understanding of microbial activity.

Proven Reserves

Proven reserves or *proved reserves, measured reserves, 1P*, often simply *Reserves* are *business or political terms* regarding fossil fuel energy sources are defined as "Quantity of energy sources estimated with reasonable certainty, from the analysis of geologic and engineering data, to be recoverable from well established or known reservoirs with the existing equipment and under the existing operating conditions." These terms relate to common fossil fuel reserves such as oil reserves, natural gas reserves, or coal reserves.

Operating conditions includes operational break-even price, regulatory and contractual approvals, of which without these items cannot be classified as proven and are usually classified into *probable*. Price changes therefore can have a large impact on classification of proven reserves. Regulatory and contractual conditions may change, and also affect proven reserves amount.

Numbered Terms

The *engineering* term P90 refers to 90 percent engineering probability, is a commonly accepted specific definition by Society of Petroleum Engineers, it does not take into account anything except technical concerns.

Therefore, it is different from the business term which does take into account current break-even profitability, and regulatory and contractual approval, but is considered a very rough equivalent. The definition is certainly not universal. Energy Watch Group uses a different definition, P95.

Indian Oil Corporation

Indian Oil Corporation Limited, or IndianOil, (BSE: 530965, NSE: IOC) is an Indian state-owned oil and gas company with its Registered Office at Mumbai, India. It is India's largest commercial enterprise, ranked 98th on the Fortune Global 500 list for 2011. IndianOil and its subsidiaries account for a 47% share in the petroleum products market, 34% share in refining capacity and 67% downstream sector pipelines capacity in India.

The IndianOil Group of Companies owns and operates 10 of India's 21 refineries with a combined refining capacity of 65.7 million metric tons per year. President of India owns 78.92% (191.62 Crore shares) in the company. In FY'11, IOCL sold 64.1 million tonnes of petroleum products and reported a PBT of Rs.9096 Crore, & Govt of india earned an excise duty of Rs.25789.90 crore and tax of Rs.1650 Crore.

IndianOil operates the largest and the widest network of fuel stations in the country, numbering about 19,463 (15,946 regular ROs & 3,517 Kissan Sewa Kendra). It has also started Auto LPG Dispensing Stations (ALDS). It supplies Indane cooking gas to over 62.4 million households through a network of 5,456 Indian distributors. In addition, IndianOil's Research and Development Centre (R&D) at Faridabad supports, develops and provides the necessary technology solutions to the operating divisions of the corporation and its customers within the country and abroad.

History

IndianOil began operation in 1959 as Indian Oil Company Ltd. The Indian Oil Corporation was formed in 1964, with the merger of Indian Refineries Ltd.

Products

Indian Oil's product range covers petrol, diesel, LPG, auto LPG, aviation turbine fuel, lubricants, naphtha, bitumen, paraffin, kerosene etc. Xtra Premium petrol, Xtra Mile diesel, Servo lubricants, Indane LPG cooking gas, Autogas LPG, IndianOil Aviation are some of its prominent brands.

Recently Indian Oil has also introduced a new business line of supplying LNG (liquefied natural gas) by cryogenic transportation. This is called "LNG at Doorstep".

Refineries

In Assam:

- Digboi Refinery, in Upper Assam, is India's oldest refinery and was commissioned in 1901. Originally a part of Assam Oil Company, it became part of IndianOil in 1981. Its original refining capacity had been 0.5 MMTPA since 1901. Modernisation project of this refinery was completed by 1996 and the refinery now has an enhanced capacity of 0.65 MMTPA.
- Guwahati Refinery, the first public sector refinery of the country, was built with Romanian collaboration and was inaugurated by Late Pt. Jawaharlal Nehru, the first Prime Minister of India, on 1 January 1962. Its capacity is 1 MMTPA.
- Bongaigaon Refinery became the eighth refinery of IndianOil after merger of Bongaigaon Refinery & Petrochemicals Limited w.e.f. 25 March 2009. It is located at Dhaligaon in Chirang district of Assam, 200 km west of Guwahati.

In Bihar:

- Barauni Refinery, in Bihar, was built in collaboration with Russia and Romania. It was commissioned in 1964 with a capacity of 1 MMTPA. Its capacity today is 6 MMTPA.

In Gujarat:

- Gujarat Refinery, at Koyali in Gujarat in Western India, is IndianOil's second largest refinery. The refinery was commissioned in 1965. It also houses the first hydrocracking unit of the country. Its present capacity is 13.70 MMTPA.

In West Bengal:

- Haldia Refinery is the only coastal refinery of the Corporation, situated 136 km downstream of Kolkata in the Purba Medinipur (East Midnapore) district. It was commissioned in 1975 with a capacity of 2.5 MMTPA, which has since been increased to 7.5 MMTPA

In Uttar Pradesh:

- Mathura Refinery was commissioned in 1982 as the sixth refinery in the fold of IndianOil and with an original capacity of 6.0 MMTPA. Located strategically between the historic cities of Delhi and Agra, the capacity of Mathura refinery was increased to 8 MMTPA.

In Haryana:

- Panipat Refinery is the seventh and largest refinery of IndianOil. The original refinery with 6 MMTPA capacity was built and commissioned in 1998. Panipat Refinery has since expanded its refining capacity to 15 MMTPA.

It is believed that the future IOCL refinery Will be Paradeep Refinery. It is expected to be handover at 2012.

International Rankings

IndianOil is the highest ranked Indian company in the Fortune 'Global 500' listing, 98th position in 2011. It is also the 18th largest petroleum company in the world and the No. 1 petroleum trading company among the National Oil Companies in the Asia-Pacific region. IOCL was featured on the 2011 Forbes Global 2000 at position 243. It is fifth most valued brand in India according to an annual survey conducted by Brand Finance and The Economic Times in 2010.

Loyalty Programs

XTRAPOWER Fleet Card Program is aimed at Large Fleet Operators. Currently it has 1 million customer base. XTRAREWARDS is a recently launched loyalty program for retail customers where customers can earn reward points on their purchases.in the org.

Oil Industry Development Board

India has begun the development of a strategic crude oil reserve sized at 37.4 million barrels (5,950,000 m^3), enough for two weeks of consumption. Petroleum stocks have been transferred from the Indian Oil Corporation (IndianOil) to the Oil Industry Development Board (OIDB). The OIDB then created the Indian Strategic Petroleum Reserves Ltd (ISPRL) to serve as the controlling government agency for the strategic reserve.

CSR activities of Indian Oil Corporation Limited

In 1964 , CSR was introduced in Indian Oil for the first time. Initially the Corporation's objective was to "help and enrich the quality of life of the community". Later on it extended its mission statement by adding " perservation of ecological balance and heritage through strong environmental conscience" to its objectives. CSR at Indian Oil is based on the four pillars of care, innovation, passion and trust. CSR here is not just to perceive as corporate behemoths that exist for profits but for the good of the society and for improving the quality of life of the community they serve. They are aware of the need to work beyond financial considerations and put in a little extra effort that can make the society a better place to live in. The Corporation respects human rights and values its employees, share holders and customers. In the last five decades , Indian Oil has supported innumerable social and community innitiatives in India and has taken up concrete actions towards the society. One of the major steps of CSR at Indian Oil is sharing of profits. Every year a fixed portion of Indian Oil's profit is set aside for community welfare and developmental programmes. It has also concerted social responsibility programmes with communities. The company has a number of community – focussed initiatives such as development in health, education, sanitation.

It also includes family welfare , providing portable water, women empowerment and welfare of schedule caste and schedule tribe beneficiaries. Allotment of petrol/diesel station dealerships and LPG distributorships to beneficiaries from among schedule caste and schedule tribes, physically handicapped, ex servicemen, war widows ,etc are some of the other initiatives under community welfare. The corporation also provides extended help to farmers to reach out to larger markets through Kissan Seva Kendras .To enforce the additional objective of preservation of ecological balance and heritage, Indian Oil has founded the Indian Oil Foundation , a non profit trust to protect, preserve and promote national heritage monuments. It is exclusively funded by Indian Oil. It has an initial corpus of twenty five crore and an annual contribution of ten crore. Under this initiative Indian Oil will adopt at least one heritage site in each state and promote it.

Indian Oil being a green company has set up some environment – protection initiatives. For the production of green fuels in its refineries it has invested nearly Rs. 7000 crore in state-of-the-art technologies. Further the corporation is now in the process of commercialising various options in alternative fuels such as ethanolblender petrol, biodiesel, hydrogen and hydrogen-CNG mixture to reduce dependence on precious petroleum products and secure the nations security.

To promote community development , Indian Oil has undertaken a number of initiatives such giving scholarships, allocation of funds etc. The community development unit of Indian Oil identifies various deserving causes for allocation of funds from the community development budget and involve various grassroot level organisations like local gram panchayats, district administration , NGOs and social workers whenever necessary.

To ensure that the benefits of the programmes flow directly to the identified groups , fund utilisation is closely monitored and whenever necessary the corporation directly executes projects. To promote all sided development of the society , Indian Oil also offers sports scholarships to upcoming players and potential/ talented players.

Indian Oil also supports art, culture , music and dance under the banners of Indian oil sangeet shabha, Indian oil art exhibitions. It has also worked towards better education in India by encouraging special scholarships among girl students and disadvantanged groups hence trying to reduce drop out rates in higher secondary levels.

Indian Oil Institute of Petroleum Management

The IndianOil Institute of Petroleum Management (IIPM) is a government run institution set up to undertake training and management development programmes for the Indian petroleum industry. IIPM has its headquarters in Gurgaon, a city in the state of Haryana, India. IIPM was set up in 1995 by Indian Oil Corporation (IOC), India's largest corporation and ranking 105th on the Fortune Global 500 listing (2009).

According to IIPM, it is an ISO 9001-2000 certificate institute, and "has been awarded the Golden Peacock National Award for 'Innovative Training Practices' by the Institute of Directors (IOD), for 1998, 2000, 2005, 2006 and for the year 2007 as well." IIPM received the "Best Innovation in Teaching" award by the Association of Indian Management Schools. IIPM also claims that it received the Indian Society of Training & Development's 'Innovative Training Practices' award for 2006-07. IIPM has institutional membership of the Delhi Library Network. IIPM officials, faculty and researchers also undertake participation in seminars at other educational institutions documenting the annual advances in petroleum refining.

Foreign Collaborations

IIPM announced in 2006 that the first pilot batch of 45 students (all employees of Indian Oil Corporation) had graduated in a joint programme with the University of British Columbia, qualifying to receive the 'Hybrid Certificate Programme in Project Management'. IIPM now plans to train over 1000 of IOC's employees under this programme.

IIPM also offers a 1-year MBA programme in Petroleum Management in collaboration with the International Centre for the Promotion of Enterprises, Ljujana, Slovenia. But IIPM allows only middle-rank officials of Public Sector Undertaking, not just from oil sector companies, to take admission in the programme.

Bibliography

Antwerpen, F.J. : *The Origins of Chemical Engineering, History of Chemical Engineering*, Washington D.C.: ACS, 1980.

Beizer, Boris: *Software Testing Techniques,* International Thomson Computer Press, Boston, MA., 2004.

Brantly, John Edward: *History of Oil Well Drilling*, Book Division, Gulf Pub. Co., Houston, 1971.

Burdick, Donald L., and William L. Leffler: *Petrochemicals in Nontechnical Language,* PennWell, Tulsa, Okla, 2001.

Chopey, Nicholas P.: *Handbook of Chemical Engineering Calculations*, McGraw-Hill, New York, 2004.

Conaway, Charles F.: *The Petroleum Industry: A Nontechnical Guide,* PennWell Pub. Co., Tulsa, OK, 1999.

Datta-Barua, Lohit: *Natural Gas Measurement and Control: A Guide for Operators and Engineers,* McGraw-Hill Professional Book Group, London, 1991.

Davis, John: *Oil and Canada-United States Relations*, National Planning Association, New York, 1959.

Diener, Daniel: *Introduction to Well Control,* Petroleum Extension Service, University of Texas at Austin, Austin, 1999.

Dunai, T.J.: *Cosmogenic Nucleides,* Cambridge University Press, U.K., 2010.

Eisner, A.: *Hydrogenation and Petrography of Some Low-rank Coals from the Western United States,* Bureau of Mines, Washington, D.C., 1940.

Frank, Alison Fleig: *Oil Empire: Visions of Prosperity in Austrian Galicia,* Harvard University Press, 2005.

Gomes, K.J., Braden, J.F.: *Mechanical Fastening of Plastics - An Engineering Handbook*, Marcel Dekker, New York, 1984.

Hill, J. W.; Petrucci, R. H.; McCreary, T. W.; Perry, S. S.: *General Chemistry*, Pearson Prentice Hall, Upper Saddle River, New Jersey, 2005.

Holler, F.J.: *Fundamentals of Analytical Chemistry,* Saunders College Publishing, New York, 1988.

Kondolf, G. Mathias; Hervé Piégay: *Tools in Fluvial Geomorphology*, Wiley, New York, 2003.

Laurence M. Harwood, Christopher J. Moody: *Experimental Organic Chemistry: Principles and Practice*, WileyBlackwell, Oxford, 1989.

March, Jerry: *Advanced Organic Chemistry: Reactions, Mechanisms, and Structure*, New York: Wiley, 1985.

Michael W. McElhinny: *The Earth's Magnetic Field,* New York: Academic Press, 1983.

Mukherjee, P.N. : *A Textbook of Objective Organic Chemistry*, Wisdom Press, Delhi, 2011.

O'Keeffe, M.; Hyde, B.G.: *Crystal Structures; I. Patterns and Symmetry*, Mineralogical Society of America, Washington, DC, 1996.

Rouhani, Fuad: *A History of OPEC,* Praeger, New York, 1971

Ruthven, D. M.: *Principles of Adsorption and Adsorption Processes,* John Wiley and Sons, New York, 1984)

Samuel, G. Robello: *Downhole Drilling Tools: Theory and Practice for Engineers and Students,* Gulf Pub., Houston, 2007.

Selby, Michael John: *Earth's Changing Surface: an Introduction to Geomorphology*, Clarendon Press, Oxford, 1985.

Suarez-Ruiz, I., and J.C. Crelling: *Applied Coal Petrology: the Role of Petrology in Coal Utilization*, Academic Press, New York, 2008.

Tissot, B., and D.H. Welte: *Petroleum Formation and Occurrence*, Springer-Verlag, New York, 1984.

Torben Smith Sørensen: *Surface Chemistry and Electrochemistry of Membranes*, CRC Press, NY, 1999.

Wachtman, John B.: *Mechanical Properties of Ceramics*, Wiley-Interscience, John Wiley & Son's, New York, 2002.

Walas, S. M.: *Phase Equilibria in Chemical Engineering,* Butterworths, Reading, MA, 1985.

Walter D.: *The Elements Beyond Uranium,* John Wiley & Sons, New York , 1990.

Wang, Y.: *Physical Chemistry and Photobiology of Nucleic Acids: Chemistry*, Athens: U of Georgia Press, 2005.

Index

❑❑❑